History, Performance and Conservation

Design professionals find, to an increasing extent, that their attention must be given to the conservation of historic buildings and environments.

This book provides, in a single text, the tools for students to evaluate such assets and to appreciate the mechanisms which can result in their deterioration. Within the context of imposed and ever-increasing standards for all buildings, the book also offers an extensive understanding of how to undertake appropriate intervention or repair and maintain the use of traditional structures without damage or loss of significance.

The book considers:

- the development of architecture within its social context chronologically, from the Ancient civilisations, through the Renaissance to modern architecture;
- the vernacular architecture of Britain, within the context of European and wider influences;
- the many different building materials utilised including thatch, slate, brick and earth;
- the history of the conservation movement and its philosophies;
- the relevance of legislation and how this is applied practically to conservation of the historic built environment.

Ideal for use in architectural technology courses, this book offers a study of the construction of our heritage and how it should be appropriately protected and conserved. It draws on a wealth of examples, offering a comprehensive picture of the architectural development of the UK within an holistic perspective and historical context, making it invaluable for both students and professionals.

Barry Bridgwood is a chartered architectural technologist. He is the compiler/author, along with Professor P. F. G. Banfill, of www.understandingconservation.org and former module leader (conservation philosophy and practice) for Building Conservation (Technology and Management) at Heriot-Watt University MSc/post-graduate diploma course.

Lindsay Lennie is a chartered surveyor with a post-graduate diploma and PhD in building conservation. Lindsay's special interest is historic retail buildings and in 2006 was awarded a Research Fellowship with Historic Scotland to survey and research Scotland's historic shops.

Technologies of Architecture
Editor: Joan Zunde

Technologies of Architecture is an introductory textbook series providing a coherent framework to the architectural design process in a practical and applied way. This series forms an essential suite of books for students of architectural technology, architecture, building surveying and construction.

Advisory board:
Michael Ashley
Mark Kennett
Stephen Pretlove
Peter Smith
Norman Wienand

Other titles:
Volume 1: *Integrated Strategies in Architecture*
Joan Zunde and Hocine Bougdah

Volume 2: *Environment, Technology and Sustainability*
Hocine Bougdah and Stephen Sharples

Volume 3: *Materials, Specification and Detailing*
Norman Wienand

Volume 4: *Practice, Management and Responsibility*
John Hickey

Technologies of Architecture
VOLUME 5

History, Performance and Conservation

Barry Bridgwood and Lindsay Lennie

Taylor & Francis
Taylor & Francis Group

LONDON AND NEW YORK

Reprint 2023

First published 2009
by Taylor & Francis
2 Park Square, Milton Park, Abingdon, Oxon OX14 4RN

Simultaneously published in the USA and Canada by Taylor & Francis
270 Madison Ave, New York, NY 10016

Taylor & Francis is an imprint of the Taylor & Francis Group, an informa business

© 2009 Barry Bridgwood and Lindsay Lennie

Typeset in Univers by Wearset Ltd, Boldon, Tyne and Wear

British Library Cataloguing in Publication Data
A catalogue record for this book is available from the British Library

Library of Congress Cataloging in Publication Data
Bridgwood, Barry, 1945–
History, performance, and conservation / Barry Bridgwood and Lindsay Lennie.
p. cm. – (Technologies of architecture ; 5)
1. Architecture–Conservation and restoration. 2. Architecture–History.
3. Building materials. I. Lennie, Lindsay Ann. II. Title.
NA105.B75 2008
720.28'8–dc22
2008016496
ISBN-9780415434201

Printed and bound in India

For sale in India, Pakistan, Nepal, Bhutan, Bangladesh and Sri Lanka only.

Acknowledgements

The authors would like to acknowledge the assistance from the following people in providing images for this publication:

- Raymond Lee
- Kevin Ward
- Elizabeth Pole
- Joan Zunde
- David Ancell

Contents

Introduction to the series

Building, architecture and technology

The categories of building, architecture and technology often, and in many ways, overlap. They can properly be distinguished. A sensible distinction is to understand architecture as a philosophical consideration of the impact of buildings on people's consciousness, while technology is concerned with the application of scientific methods to their realisation. Building has more to do with the practicalities of creating the actual structures.

The professions cannot exist in isolation, and Building Manager, Technologist and Architect, as well as all the other professions concerned with the built environment, need a grounding in each other's concerns as well as empathy with one another's points of view.

Buildings are among the most substantial indicators we have of cultures other than our own.

This is true whether we are considering varying geographical and climatic situations or taking a historical perspective. When we visit distant countries or archaeological sites, our understanding of the values and aspirations of the people who made them is vividly enhanced by our experience of the buildings we find there. They speak of the patterns of life they were built to accommodate, of the conditions under which they were created and also of the skills deployed by their designers and builders.

The buildings created today are similarly evocative. While they serve varied and complementary practical purposes, they are also markers for our sense of cultural identity. Whether we use a particular building or not, it may be a backdrop to our lives and a significant component of the environment in which we operate. It is an influential factor, whether consciously or not, in our sense of cultural identity. We should also be aware of the statement our buildings make to onlookers about our aspirations and values.

Buildings are not only the concern of those who commission and pay for them, nor of those who use them as places of work or providers of services. They are important to us all. A great hospital holds a different place in the consciousness of the Trust who own it, the medical and administrative staff who run it, the patients who use it and of the passers-by to whom it is just part of the urban scene. The same is true of the most elaborate governmental complex or of the simplest home or bus shelter.

Each of these buildings contributes to a total environment.

The village or the city is a whole formed from the constituent parts. The coherence of the experience of people within depends not only on the excellence of those individual components in themselves, seen from the point of view of owners, users or onlookers, but on the total ambience they create.

We are all, in this sense, consumers of the whole built environment.

It is the profession and art of architecture to empathise with these apprehensions of the significance of buildings, alongside ensuring that the buildings created are beautiful and practical. *Buildings that are starkly functional without relationship to their age and their place cannot be described as architecture.* Equally, edifices which simply crystallise an understanding of a culture, which stand only as features in a townscape or are merely sculptural, are follies, though possibly enjoyably decorative ones. Works of architecture serve practical purposes and do so well. They must suit the purposes of their users, must use resources wisely and must contribute positively to the visual environment.

These considerations are the concern of architectural philosophy.

The technologies of architecture, as dealt with in this series, are the developed professional skills and techniques by which the needs of the consumers of buildings in all these senses can most efficiently be met through the use of available resources. They are in every case built upon an ability to assess need, including an appreciation of what is reasonable in terms of economic, energy and time constraints. They never assume that the most modern or high-tech solution is automatically to be preferred, but always regard the low-tech and traditional as parts of the available armoury. Such technologies include aesthetics as well as acoustics, ergonomics as well as engineering and understanding both of communication and of construction.

They are, therefore, sophisticated tools which are necessary to the proper use of resources to provide an appropriate environment for the activities of society.

The expected audience

Understanding of such technologies is, of course, an essential component of the professional equipment of architects, surveyors and other practitioners, including structural, mechanical and electrical engineers as well as members of the newly emerged profession of architectural technology.

Members of all these professions need to be clearly aware of their interdependence, and work in an atmosphere of mutual respect. In some cases one or another specialist will lead the team involved in developing a design, while on other occasions he or she will be a contributing member of that team. In yet other instances a single professional may be involved.

In order to fulfil any of these roles, the practitioner needs a clear view of:

- the purpose of buildings
- the technology available to fulfil those requirements
- the specialisms that contribute to a satisfactory outcome
- how teams work
- the constraints upon the design process.

It is expected that this suite of books will be appropriate to an audience that includes students of architecture. They may be considered essential tools also for students of architectural technology, surveying and estate management and construction management in the UK, the Commonwealth and the USA.

Purpose

Volume 1, *Integrated Strategies in Architecture*, provides a preliminary examination of the knowledge, understanding and skills which the professional designer has to acquire. This text stands alone, and is written for a student without prior technical knowledge. The theoretical topics covered are fundamental and basic, and are introduced by way of material with which the reader may be expected to be familiar.

Volume 2, *Environment, Technology and Sustainability* deals with the management of the physical environment, while Volume 3, *Materials, Specification and Detailing* is concerned with quality issues and communication of solutions, and Volume 4, *Practice, Management and Responsibility*, will approach questions arising from the business of architectural practice.

The present book, Volume 5, takes the practitioner into the area of care for the existing built environment and concern for its future via an understanding of its historical development, use of materials and after-care led by a developed philosophy and practice of conservation.

The whole suite of books is conceived as a set of course texts rather than as reference materials, since the breadth of data that would be necessary for such books is beyond the scope of student manuals.

Part 1
Introduction

Part **1**

Introduction

This book is part of a series of volumes which are investigating the art and science of architecture in its widest sense. This volume is concerned with three main subjects: architectural history, materials technology and building conservation. These three elements are closely intertwined and while each may stand on their own, it is only from an understanding of all three that it is possible to be in a position to act as a conservation practitioner. Understanding of how buildings were designed, what inspired their styles and the materials they were created from all form the foundation for protecting those buildings. This clarity of understanding then allows informed decisions to be made about the future of the historic environment.

Chapter **1**

Architectural history and materials

The study of architectural history is a basis for understanding historic buildings and for implementing their conservation and protection. The relevance of examining historical development is outlined by Conway and Roenisch (2005) who state that "studying the past can help us understand how we have arrived at today and give us insights into the production and use of built environments". Further, Furneaux Jordan (1997) considers the different elements which contribute to architecture stating that:

> the ... structure and forms of architecture were almost always the product of time and space – of circumstance more than will. Man's thoughts and actions – his religion, politics, art, technology and aspirations, as well as landscape, geology and climate are the things from which an architecture is born. The art of a civilization, rightly interpreted, is a very precise reflection of the society which produced it.

The earliest human built environments involved the adaptation of natural features like caves or the use of locally available materials in order to provide shelter from the elements and defence against predators or enemies. The available resources would probably have been limited to the local environment, so the beginnings of shelter were determined by geology, geography and the availability of suitable materials. However, such shelters were eventually replaced with more permanent societies as communities moved from their simple buildings and adopted the essentials of community, cultural enhancement and decoration. This would not have occurred at one particular point in time, but instead happened gradually, as a continuum, and would have also varied from one society to another depending on their particular local circumstances.

The influences on the evolution of architecture are numerous but are well defined in Banister Fletcher's (1928) *Tree of Architecture*

(Figure 1.1). In this diagram he clearly outlines the various factors which shape architecture. At the roots of his tree are the fundamental determinants:

1   geography
2   geology
3   climate
4   religion [or may be termed ethics]
5   social and political
6   history.

The branches and canopy of the tree are formed from the various architectural styles, the lower branches representing the most ancient styles and the upper canopy representing the more recent. These styles are identified as follows:

1   Peruvian, Egyptian, Assyrian and Chinese/Japanese as the earli-
    est, followed by
2   Mexican, Greek and Indian – these were later developed but were
    not directly influential (with the exception of Greek) in
3   Roman, which absorbed some of the former influences (primarily
    Greek and Middle Eastern), developed them and were later fol-
    lowed by
4   Byzantine, Romanesque and Saracenic followed and enlarged on by
5   locally stylised, medieval gothic, followed by
6   Renaissance architecture and its later developments, swayed by
    Palladio, Vitruvius et al. and extending into Neo-Classical forms
    (influenced by the Enlightenment movement) during the late
    eighteenth and early nineteenth centuries.

The more recent periods of architectural development might be added and are simply categorised, in the modern western world as:

*   Industrial (nineteenth century)
*   Neo-gothic or Gothic revival
*   Arts and Crafts and Art Nouveau
*   Art Deco
*   Modernism
*   Post modernism
*   Contemporary
*   Internationalism.

This pattern and continuum of cause, effect and response is based on experience, understanding of influence, plagiarism and recognition. It must also be remembered though that these styles will be adapted by particular local circumstances. However, this "tree" provides a good basic structure for understanding the chronology and development of architectural styles.

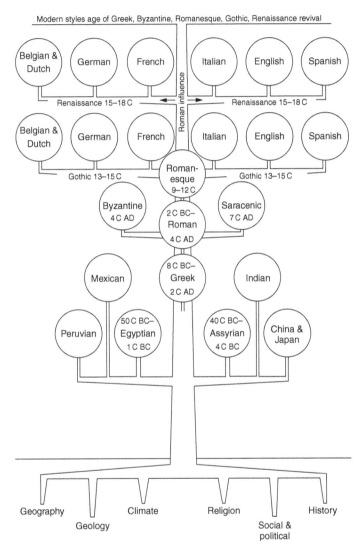

Modern styles age of Greek, Byzantine, Romanesque, Gothic, Renaissance revival

Belgian & Dutch | German | French | Roman influence | Italian | English | Spanish

Renaissance 15–18 C — Renaissance 15–18 C

Belgian & Dutch | German | French | Italian | English | Spanish

Gothic 13–15 C — Roman-esque 9–12 C — Gothic 13–15 C

Byzantine 4 C AD | 2 C BC– Roman 4 C AD | Saracenic 7 C AD

Mexican | 8 C BC– Greek 2 C AD | Indian

Peruvian | 50 C BC– Egyptian 1 C BC | 40 C BC– Assyrian 4 C BC | China & Japan

Geography   Climate   Religion   History
Geology   Social & political

1.1   Drawing of Tree of Architecture (after Banister Fletcher)

In the early sections of this volume we will look, in greater detail, at these stylistic influences, how they have formed our built environment and why we need to be able to read that developmental story. This allows us to evaluate what is important so that repairs, maintenance and alterations to historic buildings are carried out appropriately or offer realistic and viable alternative uses without damaging historic, spiritual or aesthetic value.

Part of that process of understanding involves a consideration of the motivations for the creation of the built environment. Why we build beyond our basic needs and why, as part of the development of the built environment, we transform buildings into art and architecture. This is central to understanding architectural history, together with an appreciation of chronological development. This book will therefore consider the historical development of building styles, from ancient civilisations to the present day, as a context for conservation philosophy. However, as many of these areas of architecture are specialist subjects in their own right, the reader will be referred to suitable texts for further reading on these topics.

Having established the foundation of architectural history as the basis for architecture and conservation, it is also essential to appreciate the various materials that are involved in building technology. The use of particular building materials can have a significant impact on the potential longevity of the historic and current built environment. For the sake of style and appearance and perhaps through a lack of understanding of materials, buildings in the past were sometimes erected with inherent defects. For example, some Georgian brick buildings have a double skin of brick which is not tied together. We therefore need to understand these potential defects in order to make appropriate repairs without losing historical significance. A failure to understand these buildings can result in unnecessary and perhaps damaging interventions.

The adoption of materials for the construction of buildings is a complex series of cause and response; what is available, what is economic, what is needed and what is appropriate. The fundamental influence within this process is communication. We are affected by what we know, how we perceive it and how we are informed about it. In earlier times society was subject to a very narrow band of influence based on what was available locally and how far we might travel in a day. What is available might not just relate to materials but also to more esoteric values such as how we perceive ourselves and what value systems that generates. Architecture reflects, rather than dictates society; it is still, however, vicariously influential. Once that truth is recognised then the value of the built environment becomes easier to assimilate and understand.

Chapter 2

Conservation and the built environment

In the English Heritage publication *History Matters: Pass it On* (2006) the importance of Britain's historic environment is clearly set out as follows:

> history matters ... a society out of touch with its past cannot have confidence in its future. History defines, educates and inspires us and a better understanding of our shared history helps us solve the problems we face.... Our heritage – our historic buildings, landscape, archaeology and gardens, and the art, books and machines that they contain – is a set of essential landmarks to guide our journey from the past to the future.

The built environment is extremely complex yet may be read, by those experienced in doing so, as we would read a book or examine a piece of artwork. Learning how to read buildings is a crucial step for any professional in conservation. For example, major historical events such as the wholesale desecration of the English religious environment during the Dissolution of the Monasteries in the mid-sixteenth century or, more recently, the Second World War, are reflections of the social, economic and political circumstances of the time. Buildings reflect those events, demonstrating both influence and response. These may overtly demonstrate a period, such as war-time bunkers dating to the Second World War, or more subtly as in the loss of statues in ecclesiastical buildings during the Reformation.

This myriad of structures contributes to the continuum of architectural history. All aspects of the historic environment should be considered for protection in order that all aspects of society are fairly reflected. It is particularly important not to be judgemental about the architecture of previous societies. This is aptly described by Osbert Lancaster (in Earl, 1997) who states: "Let us always be aware of the uncertainty of private judgement, remembering that what to us may be without merit may well prove to posterity, who can view it in perspective, of considerable value."

William Morris, in the Society for the Protection of Ancient Buildings' Manifesto of 1877 also strongly defended the need to protect historic buildings, stating that:

> These buildings do not belong to us ... they have belonged to our forefathers and they will belong to our descendants unless we play them false. They are not ... our property, to do as we like with. We are only trustees for those that come after us.

Both quotations encapsulate the importance of appreciating that current generations are merely stewards of the built environment. We therefore have a responsibility to pass on that environment in a way that neither reduces the value of the past nor affects the ability of present and future generations to form their own view, impose their own values and in so doing maintain an accurate record of their society and make their own assessment of value.

The lack of respect for more recent buildings may be partially attributed to the rapid technological change which society has experienced in recent decades. New technology such as the Internet has provided us with such a vast source of information that it may confuse rather than clarify. This rapidity of change, in addition to potentially confusing us, has encouraged a "throw away society" that may pay less regard to what we are achieving now simply because the past may seem more comfortable and "value-some" than the present. Too rapid a rate of change makes us uncomfortable because we long for the values of "old fashioned society" when things happened at a less frenetic pace and life was, by perception, easier and less pressured. The hazard in this process is that we place less value on the aspects which represent our current society and instead there is a tendency to prefer the past.

We therefore need to recognise that what may seem to us to be of little value, will form the history of succeeding generations and in doing so may acquire values that we do not presently recognise. The current vogue for "retro" styles such as the 1950s, 1960s or 1970s in clothing, household items and television programmes demonstrates how quickly relatively recent history – which was not so long ago ridiculed – is now revered. The same also applies to buildings. Victorians had little regard for the buildings of their Georgian predecessors. In the twentieth century many Victorian buildings were demolished as being of little historic importance. The demolition of even more recent buildings such as the 1930s Art Deco Firestone Factory galvanised support for modern buildings, largely encouraged and highlighted by groups like the Twentieth Century Society (formerly the Thirties Society). The more recent the building, the harder it is for society to accept its importance and possibly its place in history. The controversy over the cost and delayed completion of the Scottish Parliament building has clouded judgement over its design and it will be for future generations to judge whether the significant cost has created a building of historical significance. Other examples of modern buildings such as St Mary Axe or Canary Wharf may not be currently appreciated, but a future generation may come to regard them as significant.

In approaching conservation, it is vital that we build on previous experiences and learn lessons from the past so that any interventions are appropriate. Conservation philosophy encourages respect for what has gone before but, at the same time, not losing sight of the value to future generations that our contemporary society has to offer. A balance must be achieved between the past, the present and the future if we are to appreciate contemporary influence. We must respect the ability of future society's ability to form its own judgement about value without imposing our own, possibly subjective, views.

Chapter **3**

Managing change

While conservation of the built environment is central to this book, in the context of architectural history it is equally recognised that buildings are not museums and should not be fossilised. While some very special examples may be maintained in a "preserved" state, the majority of buildings need to have a working purpose and be practical for the needs of their occupants. Being suitable for particular purposes may require a building to be altered, but how that alteration is managed is crucial. Fielden (in Earl, 1997) states: "Conservation is very largely the art of controlling [or managing] change."

Change must be managed without compromising the value or significance of the built heritage, while facilitating its effective and economic use. Without change historic buildings may not have a secure and viable future but that change must be both considered and appropriate for that building.

Decisions must be taken in the full knowledge of the history of the building, its materials and construction methods. From this research develops an understanding that will then underpin any decision-making process. Such a process can be complex and may involve many different facets of investigation. For some buildings the answer may not be obvious or easy. For example, for numerous redundant churches the options for re-use may be extremely limited. In other cases, owners may want to dramatically change the building, such as the restoration of a ruined castle to make it into a habitable house. These pose philosophical and economic dilemmas and considerable research and discussion is likely to be required before the problem will be resolved.

For other buildings, trying to make them conform to modern requirements can be a significant issue. It is important to recognise that modern value systems differ greatly from those in place when traditional buildings were being constructed. Modern impacts include legislation, performance and use. We try to make historic buildings incorporate floor loadings, services, access requirements and building standards that they were never designed to accommodate. In adapting historic buildings it is vital that we do not force buildings to carry out a function for which they were never designed and cannot realistically achieve. Instead, a compatible and sympathetic re-use must be found. In this way we can respect a building's history and significance and only require it to provide what it is capable of achieving without losing any significance. However, in certain cases with very specialist buildings, such as the O_2 Dome (formerly the Millennium Dome), it has to be accepted that this may be limited. It may take a leap of faith in some cases, such as the use of the former Battersea Power Station for the Tate Modern Art Gallery.

Chapter **4**

Conclusion

History is a continuum. It is affected by many closely inter-related factors, philosophical, physical and spiritual. It offers a palimpsest (a manuscript which is written and then written over) which needs to be understood, maintained, conserved and protected. There should therefore not be any interventions to a historic building without first establishing if our comprehension of the building and its development is accurate.

The subject of architectural history is a considerable one spanning several millennia. However, an understanding of architectural development informs conservation philosophy and is the foundation, together with understanding of materials technology, for building conservation.

Part 2
Early architectural history

Part 2

Introduction

The study of architectural history is fascinating and has been pursued by scholars over centuries. From such study emerges a deep understanding of buildings; how they were formed, where they were sourced and what inspired them. Over time architects have revisited the past and adapted and re-used the design ideas of their predecessors. It is therefore crucial to understand the sources of these ideas.

This and the subsequent part will follow a chronological approach, though this can never be exact as time periods overlap and cannot be precisely defined. Although the focus is on the architectural history of the United Kingdom, the influences of cultures from Europe and beyond will also be examined.

To gain an initial understanding of architectural history we should take a short sojourn outside Western Europe and look at ancient civilisations and their cultural influences on architecture. This section of the book is not by any means intended as a definitive analysis of ancient architectural styles and influence, its purpose is merely to offer a sketch of architectural development. This will include Egyptian, Western Asiatic, Greek and Ancient Roman forms in order to demonstrate the various influences of their cultures and how they ultimately affected subsequent cultures.

The change from small family or tribe-based collectives probably emerged during the transition from simple hunter-gatherer groups to more cohesive social interaction into those that might be identified as communities. The primary motivator was the move from a hunter-gatherer-based society to one based on the simple cultivation of crops. The location of water was therefore crucial in this process, both as a source or drinking water, then as crop irrigation and much later as a means of communication and travel. The ancient civilisations of the Nile Delta in Egypt, the Yangste in China, and in Mesopotamia round the Euphrates and Tigris, typify this development.

So, we see that civilisations started, not with any central influence, but more around locally available water sources across disparate geographical areas and sites. This resulted in geographically dispersed, independent cultures with their own approach to design, art and architecture that reflected their simple needs and requirements. They utilised locally available materials but also had an aesthetic dimension, despite their limited field of influence. So, it might be proposed that early civilisations are the only ones that developed with an introverted focus and were not influenced, in the early stages at least, by any outside culture.

Architecture within these isolated civilisations was very much based on what materials were available locally and how they might be used, adopted and adapted to form shelter; to make statements about their culture and allow the community to exist within an environment dictated by need, weather, protection and, eventually, defence.

Chapter **5**

Western Asiatic, Assyrian and Persian-Mesopotamia (from 4500 BC)

The ancient civilisations of Mesopotamia, Persia and Assyria were located in what we now know as the Middle East. Mesopotamia, referred to as the "cradle of civilisation", had a complex and turbulent history. The period lasted from about 4500 BC and key historical events include the invasion of Egypt by the Assyrians in 672 BC and the destruction of Ninevah in 609 BC. Babylon was conquered by the Persians in 539 BC and the Persian Empire witnessed a period of palace building, the most important being at Susa and Persepolis. The country remained under the rule of Persia until 333 BC when Alexander the Great took it as a possession of Hellenic Greece (Fletcher, 1928).

The people of this area were extremely superstitious, worshipping the sun, the moon and natural forces. They tended to celebrate their religious practices in the open air and so large-scale religious buildings were not required, although they did construct grand palaces. Aggressively war-like, they used prisoners captured during their expansive escapades to provide the labour force for the construction of their major architectural edifices.

Ancient Mesopotamian society may claim to have constructed the first identifiable large town or city and developed a community-based culture centred on agriculture well before the Ancient Egyptians. This sophisticated society is believed to have invented the wheel (tripartite solid wheel), they wrote in the form of cuneiform symbols embedded in clay tablets and developed a system of agriculture that supported its society. The Mesopotamians were thought to have developed the first plough and, along with bread produced from the grain grown using systematic agricultural methods, also produced beer from barley.

Mesopotamian and Western Asiatic architecture is best defined by the use and development of mud and clay, including sun-dried bricks, kiln-dried bricks and glazed bricks. Early Western Asiatic buildings made extensive use of mud structures and sun-dried bricks. This later developed into the use of kiln-burnt bricks to improve longevity. Bearing in mind that the geographical area defined by the Euphrates and Tigris rivers, now in modern-day Iraq, was a heavily cultivated region with no endemic stone but abundant alluvial muds and clays it is hardly surprising that its building forms made use of mud and clay bricks.

In Assyria, located north of the Euphrates and Tigris rivers, naturally occurring stone allowed for the development of a stone-faced built form with brick structures covered both internally and externally with alabaster and/or limestone slabs carved into flat or low (basso-rilievo) relief. These low or bas-relief carvings provide a very detailed picture of life in Assyrian culture and, like so much of our built environment, if properly read with an understanding of language and significance, provides a palimpsest of history to which it bears witness.

The use of the arch and vault, resting on thick and heavy walls, is thought to have originated in this area. Bear in mind that the Mesopotamians rarely used stone as a constructional form in its own right, relying on it only to provide decoration by applied panels. The

5.1   Drawing of basso-rilievo relief

Persians of the north, however, did adopt stone in structural and supportive forms as walls and isolated supports in the form of columns with attendant arches, vaults and cupolas. It is considered (Fletcher, 1928) that the forms adopted by the builders of these structures took much from original timber methods of construction and echoed such details within stone equivalents. This is, perhaps, resonant of the Egyptian tactic of adopting earlier reed shapes for their columns, this, in turn, reflecting former construction practices. Eventually, this use of timber construction became superseded by the advancing ability of the artisans (stonemasons) by developing a form that was more applicable to this new material. Stone therefore dictated architecture and took on its own format and approach. This is exemplified by later Greek builders and architects and their adoption of exceptionally accurate stone cutting; underpinned by a clear comprehension and use of mathematics which assisted in the calculation and design of buildings.

There is some speculation that an early form of the Doric and Ionic Orders originated in Persia, with the earliest form of Doric being seen in Egyptian columnar design at Beni-Hassan rock tombs, well before the Greek forms. Another example, perhaps, of previous cultures influencing following societies.

5.2   Drawing of typical Assyrian architecture

Chapter 6

Ancient Egypt (from 5000 BC)

> The Egyptian pyramids have survived thousands of years, but historical significance is not just a question of durability. These buildings were part of a rich and diverse culture.... They are historical facts, but facts by themselves ... are just the first stage in any historical study, and until they have been placed in context and interpreted, they tell us very little.
>
> Conway and Roenisch, 2005

The culture of the Ancient Egyptian civilisations, centred round the Nile Delta and Upper Nile, was focused on a need to allow the human body to move from its earthly existence to an afterlife. At the centre of that belief system was a need to preserve the human body by mummification, preservation and protection in order to facilitate a transition to a life after death. This led to the construction of tombs to protect the mummified remains and to allow the spirit to transit from its earthly existence to an afterlife, protected against the effects of decay.

The early tombs (mastaba) were simply mounds, made up of large rectangular blocks of stone laid one on top of the other in pyramid form. Tombs were also repositories for the accrued wealth of the individual for enjoyment in the afterlife. Such riches needed protection against plunder so the architecture of the impregnable tomb became an integral part of the Egyptian culture. Starting from these simplified roots, eventually size – to reflect status – started to influence tomb structures. The larger the tomb the more important the person buried within it. The pyramids at Giza were the culmination of tomb development, each one being larger than its former to reflect the status and power of the succeeding pharaoh. They were erected 3733–3566 BC.

Tomb structures were essentially mortuary buildings – repositories for the remains of the mummified pharaohs and a celebration of their

6.1   The Great Pyramids at Giza

6.2 Drawing of Queen Hatshepsut temple at Dier-el-Bahari

lives. They reflected various cultural elements including status, development of mathematics and the technology of using stone for monumental construction. So, we might say that this was architecture in its earliest development. The construction of Egyptian temples clearly demonstrates the development of culture through building and its elevation into an art form expressive of that culture. Their monumental size reflected the power of the state, its religion and deification of its leaders, the pharaohs. The mammoth task of creation of temples such as that at Dier-el-Bahari (a monument to Queen Hatshepsut) dating 1550–1330 BC, set against the backdrop of the mountainous rocky cliff behind it, or that of the temples of Rameses II at Abu Simbel clearly demonstrates the establishment of architecture within the context of landscape.

The pyramids and mortuary temples exemplify the ingenuity of the Ancient Egyptians and their particular mastery of the craft of stonemasonry. This, perhaps, was the precursor to the liaison between religion and celebration of faiths and the craft of the stonemason later reflected in Greek temple buildings and in the great cathedral-building era in medieval Western Europe. Certainly, the later Egyptian architects who created such edifices offered a foundation which encouraged the development of architecture as an art form. Eventually this approach was adopted and synthesised by superseding cultures.

6.3 Temple of Hathor, Dendera, showing trabeated form of architecture

The Ancient Egyptians had no knowledge of the arch. Their architecture and structures were therefore based on a trabeated form of construction where the columns support the beams or lintel. This is different from the arch-based architecture utilised much later by the Romans through their development of the semi-circular arch. However, it is recognised that Assyrian architecture may have made use of arches, vaults and domes. This trabeated form had a number of limitations. Primarily it required many columns for support and it did not facilitate the creation of large enclosed spaces for assembly.

The Egyptian religious practices limited entry to the inner sanctum of the temples to priests and the king or pharaoh and his queen, so large public assembly spaces were not required. Rhetorically, did the layout of these buildings influence the form of worship or was it the other way around? So, post-pyramids Egyptian architecture is dominated by column-supported flat-topped temple buildings. Such column-dictated buildings are called hypostyle halls; often with raised central bays to provide clerestory lights to the depth of an otherwise unlit interior. These buildings were heavily decorated with hieroglyphs incised into the surface of the stonework. These provide a record of the lives of the Egyptians and their pharaonic society.

The architecture of the Ancient Egyptians was inspired by their natural surroundings, notably the papyrus and lotus plants, and this became embodied in Egyptian columnar designs. The use of papyrus reed

6.4   Columns of a hypostyle hall at Karnak, Egypt

6.5   Temple at Edfu, showing incised hieroglyphs

influenced columnar form and possibly originates in the Egyptians' earlier use of mud buildings which were constructed using papyrus stems bound together as an initial framework for creation of housing. None of these buildings survive. However, the necking of Egyptian columns at the base of the capital is reflective of a ligature around the head of plant stems when bound together to form a building element. If this assertion is correct, then it may be considered to be an example of historic practices being subsumed within a built form that no longer uses earlier techniques but is influenced by them – the impact of history and of borrowing from previous practices and methods.

Egyptian buildings may date from as early as 4777 BC. Egypt became dominated by foreign influence during the period 950–663 BC, so Ancient Egyptian civilisation and culture lasted for a minimum of 4,000 years. What is it about Ancient Egyptian buildings that they have managed to last almost three millennia beyond their last great period of building? Possibly the method of construction but also probably the material used in the construction of their edifices. The Ancient Egyptians had superb stone sources in the limestone of the north, sandstone in the central belt and granite in the south. The granite sources found largely in Aswan are the reason why so many buildings remain available for study today.

6.6   Plant-influenced column forms

Egyptian architecture is one of the main architectural foundations on which subsequent designs were based. It influenced Greek and Roman and is evident in more recent architectural styles such as Art Deco and Neo-Classicism. This is therefore a clear example of later cultures adopting the designs of former, and sometimes ancient, civilisations.

Chapter **7**

Ancient Greece (1100 BC–146 BC)

In describing the importance of Greek civilisation, Banister Fletcher (1928) states that they "by reason of their innate artistic sense, so profoundly influenced the development of European art that Greece must be regarded as the veritable source of literary and artistic inspiration". An understanding of this complex and ancient culture is an essential tool in comprehending later architectural development.

The Greek land mass differs from that of Egypt in that it is dominated by rocky inlets, islands and mountainous regions, making early communication across the country difficult. As a result, communities tended to evolve as independent entities, although the development of a maritime culture did overcome some of these difficulties. Like Egypt though, the Greek civilisation dates back many thousands of years to the Neolithic period, although the main periods of interest are identified as follows:

- Ancient: 1100–750 BC
- Archaic: 750–500 BC
- Classical: 500–336 BC
- Hellenic: 336–146 BC.

Two tribes, the Ionian and the Dorian, dominated Greek culture during the Ancient era indicating the origins of the terminology for two most well-recognised and used Greek "Orders" of Ionic and Doric. During the Archaic period, art and culture began to develop but it was in the Classical period that culture, philosophy and democracy were fully established in Greek culture.

The Greek civilisation was a highly developed culture. The great philosophers, Plato, Socrates and Aristotle used open public spaces and markets (agora) as meeting and debating places and developed a culture of theatre. If we consider the Egyptians as sculptors, via simple two-dimensional hieroglyphs the Greek sculptors made their statues come to life: "it is about to breathe, not because it is realistic but because it is instinct with life" (Furneaux Jordan, 1997). The Greeks were analytical, investigative and great mathematicians. They developed a democratic system but with an educated, philosophical, oligarchic elite in positions of power.

Above all, the Greeks were artists and this is reflected in their architecture. They used their knowledge of mathematics to seek perfection in their buildings and they were constantly striving to achieve beauty. They admired the body beautiful, fitness and great physical attributes and these were reflected in their statuary. Their development of the Games reflected these aspirations.

Cities were based around tribal foci and for defence at Athens, Sparta and Corinth, although Athens became the central focus of Greek society. The architecture of their theatres or amphitheatres provided little for the development of an architectural style and, similarly, their houses were of limited architectural interest. However, it is in their temples that we see significant advances in architectural development, such as the Athenian Acropolis containing the Parthenon, built 454–438 BC.

7.1   Parthenon, Athens

Built on high ground above a city, these citadels had gravitas and importance, particularly evident when they were silhouetted against the skyline. The subtle interplay of light through a structure provided the Greek architects with a canvas to demonstrate their mastery of the art and science of architecture. Their seeking of perfection was reinforced by their ability to control the play of light against dark in very subtle ways.

They attempted to achieve 'perfect' buildings through the use of mathematical techniques. This included placing the columns at the corners of the building fractionally closer together in order to counter the optical illusion of those columns apparently being further apart because of the effect of light around them. In addition, columns were angled fractionally inwards so that, if extended vertically, their axes would meet two miles above the ground. The Greeks also used entasis, a barely perceptible outward bulging curve on the sides of columns and other structural elements, to counter the illusion of distortion through perspective.

Greek masonry was carved and worked with precision; no binder was used, the separate stones were ground together with microscopic exactitude to ensure a perfect fit. In this way they developed the art of the stonemason to a science, utilising readily available marble to great effect. Their buildings were sculpted to produce a built environment that was a combination of science and art but with the seeking of artistic perfection as the dominant influence.

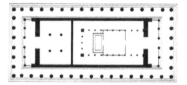

Although the Greeks were aware of the arch and the vault, they did not develop it, preferring a trabeated form of architecture so epitomised by the Classical Greek temple and reflective of the Egyptian hypostyle hall.

7.2   Drawing of a typical Greek temple

7.3   Typical carvings found on tympanum

The classic Greek temple was rectangular in plan with a range of columns, or peristyle, surrounding it. The roofs were low-pitched with triangular gables or tympanum, usually heavily carved with mythical scenes, such as the birth of Athena on the east tympanum of the Parthenon.

The columns of the peristyle were surmounted by an entablature, usually in three layers: the lowest being the architrave, the mid-section known as the frieze (sometimes carved with scenes or motifs) and above it the cornice. This is a very simple description of entablature and variations are demonstrated in the separate orders of Doric, Ionic and Corinthian.

The temple structure, comprising the peristyle, entablature, tympanum, roof and temple enclosure, was placed on a stepped approach platform known as a crepidoma, the top step of which is called the stylobate. The whole structure was designed to be viewed from afar and from below, hence the choice of elevated sites divorced from its supportive city, perhaps related to the idea of being the homes of the gods in the sky.

The Greeks adopted three "Orders" or styles for their columns, Doric, Ionic and Corinthian. Summerson (1996: 19) defines the Orders as "columns supported on pedestals (whose use is optional) and carrying beams with projections to support the eaves of a roof". These are central to Greek architecture and the ideas were subsequently adopted and developed by the Romans. Although they were developed by the Greeks, it may be that they had their genesis in Persian and Egyptian forms. Alexander the Great (356–323 BC) conquered Persia in 333 BC and then Egypt. Subsumed within that conquest there may be a reverse influence of absorption via conquest, imposition and subjugation. Conquerors may impose their own ideas, but they also absorb some of the ideas of the conquered nation, resulting in cross-cultural influence.

Whatever their origins, the following are very simple definitions of the three Greek Orders and the reader is encouraged to investigate examples in order to further understanding by studying examples and by reference to the recommended reading list. For the purposes of clarification, Orders include the column and base and the entablature or supported part.

The Doric Order is strong and powerful with thick, masculine columns which are usually fluted. The columns have no base or pedestal and simply start at the base of the structure above the stylobate. They terminate in a very simple capital with a circular echinus ("sea urchin") and flat square abacus. The architrave is equally simple and unadorned and is surmounted by a frieze, usually made up of triglyph below both the tympanum and expressed within the frieze. The

7.4   Drawing of Beni-Hassan rock tombs in Egypt showing an early form of Doric-style columns

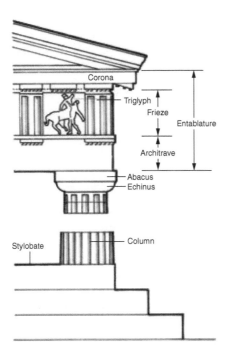

Corona

Triglyph

Frieze

Entablature

Architrave

Abacus

Echinus

Stylobate

Column

7.5 Illustration of Doric Order

spaces between the triglyph are usually carved panels or metope. Above the frieze is a cornice, very simple in design, reflecting that of the rest of the Order.

The Ionic form develops the Doric Order but achieves a more sophisticated or subtle ensemble. The slender columns are still fluted but spring from a defined attic base or pedestal; the capital is more ornate with scroll-like termination (volutes – parallel, as opposed to the corner-projecting Roman version). The entablature is again subtle and less heavy in appearance than the Doric style. The frieze may be carved or uncarved, but the cornice is sometimes terminated in a double curved section (Cyma reversa/Cyma rectus) separated by a flat square corona, sometimes underpinned by modillions.

The Corinthian Order is a much more decorated style, more ebullient and expressive than the previous two Orders. The columns are fluted or plain and spring from a decorated base or pedestal with a plinth. The capital is heavily decorated and articulated with acanthus leaves and floral volutes springing from each corner; these are usually pointed or chamfered. The entablature is invariably plain and unarticulated: first there are three fasciae below a frieze topped by a dentil

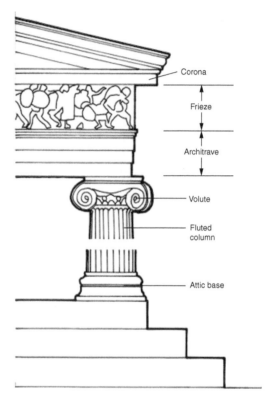

7.6   Illustration of Ionic Order

course and an overhanging cornice, probably with antefixa above at the edges.

The period of greatest development and expansion of Classical Greek influence was during the Athenian period, probably limited to two or so generations during the period 500–400 BC, under the influence of Pericles. Its subsequent decline, like its ascendance, was extended over an equally long period. Furneaux Jordan (1997) likens it to "the perihelion of a comet – a long slow preparation ... a short blaze of achievement, then the long, slow decline". Nonetheless, Greek art and architecture undoubtedly had an extremely extended influence on later societies across many centuries and numerous geographical boundaries. Notably, Greek architecture was much admired and was

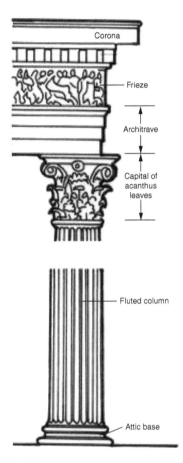

Corona

Frieze

Architrave

Capital of
acanthus
leaves

Fluted column

Attic base

7.7   Illustration of Corinthian Order

further developed by the Romans. It also significantly influenced later Western European architectural periods including the Renaissance and the Neo-Classical revivals of the eighteenth century. It is echoed in numerous stylistic architectural responses and is still studied and revered today.

Chapter **8**

Ancient Rome: 750 BC to AD 400

While the Greek period was temporal and aesthete, Roman architecture may be regarded as secular, expansive and demonstrative of power and status. Above all, Rome was dismissive of Greek effeminacy and trickery but was admiring of its achievements in philosophical thought, artistry and subtlety. Rome absorbed the Greek styles and adapted them to suit its own purpose. Rome extended and developed the art of the Greek architect but used it to achieve a more secular, perhaps less artistic, focus which demonstrated power and subjugation of those it conquered and vanquished. If the Greek was art through architecture then the Roman was architecture to demonstrate power and status.

In terms of religion, the Ancient Romans worshipped many gods and deities, following a pantheistic form of belief. Indeed the most well-known of all the Roman temple buildings, the Pantheon, is a place of worship of all the Roman gods, the name meaning an ancient temple dedicated to all the gods.

The Romans were expansive and conquering, with an Empire which spread across much of the known Western world. At its height it extended from Britannia (Britain) in the north to Greece and Western Asia in the east and south and into northern Africa. The reign lasted for approximately 1,100 years from about 750 BC to AD 400. The Empire expanded hugely under Hadrian (117–138 BC) and the lattermost emperors, Caracalla (211–217 BC) and Diocletian (284–305 BC) did most to demonstrate the architectural abilities of Ancient Rome. Rome descended into decay from about 305 BC until Constantine (306–337 BC) revived some of its influence under Christianity. Rome was sacked by Alaric in AD 410 and had been subject to Teutonic invasion during AD 376.

So, what *did* the Romans do for us? They certainly gave us the earliest form of lime-based concrete or pozzolanic concrete, they gave us the

8.1   Pantheon, Rome

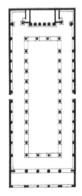

8.2   Typical plan form of Roman basilica

semi-circular arch and the dome, the Pantheon in Rome being an example of the largest unsupported dome anywhere in the world at that time, its construction facilitated by the use of concrete. They also provided us with the basic plan form of our own ancient cathedrals, the basilica. Roman basilica were buildings for the suit of law, very similar to our own court buildings.

The Romans certainly admired the Greeks and Ancient Egyptians; they imported works of art and principles of architecture from both cultures. They adapted the Greek Orders and made them their own, developing them to an extended series of five. In that process they might be accused of plagiarism in the raw, especially in respect of Greek architecture. They were, after all, primarily conquerors rather than artists. This is not to degrade their artistic talents and abilities but more to recognise that their cultural focus was, perhaps, less than the Greeks' in terms of pure philosophy and thought in preference to a more secular culture. They were, however, equally influential culturally as they were politically, and numerous cultures benefited from their influence, architectural and artistic styles. The temples at Petra in Jordan (dating to AD 150) very typically demonstrate this Roman influence in areas where Rome conquered and developed cultures and styles.

The five Roman orders of architecture are:

1   Tuscan
2   Doric
3   Ionic
4   Corinthian
5   Composite.

8.3   Petra Treasury, Jordan

By simple comparison with the Greek Orders, the Roman might be defined as more ornate, possibly with the exception of Tuscan which is particularly austere and void of excess of decoration or ornamentation. The following are simplified descriptions of the five Roman Orders:

Tuscan is the simplest and least decorated. The columns are typically plain over a simple square base with torus, fillet and apophyge. The column terminates in a simple capital with apophyge, fillet, astragal, echinus and abacus. The entablature is again simple, comprising of an architrave consisting of two fascias with a frieze over, surmounted by a cornice.

The Roman Doric Order is similar to Greek Doric but with a greater degree of decoration at the column head. As opposed to the Greek, the column has an attic base, but is fluted and has a capital decorated with egg-and-dart enrichment. The entablature has a simple fascia below a frieze with triglyph and pelta, surmounted by a cornice with mutule.

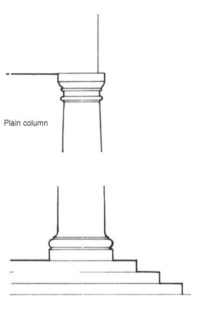

Plain column

8.4   Illustration of Tuscan Order

The Ionic Order has column shafts that are invariably unfluted and are usually provided with an attic base. The simple parallel volutes of the Greek form (with its need for special corner columns) are developed in the Roman form by allowing the volutes to "radiate" from the corners of the abacus. The entablature is simple with architrave, frieze and cornice.

The Corinthian Order is highly decorated with plain or fluted columns and heavily carved acanthus leaves in the capital. The entablature is quite deep with three fascia surmounted by a narrow architrave or cornice, with a plain frieze over-topped by a cornice with egg-and-dart and modillion decoration.

Composite is the grandest of the Roman Orders. It has an ornate version of the Ionic capital with radiating volutes above two carved rows of acanthus leaves. The entablature is usually heavily decorated and is similar in appearance to the Corinthian.

An understanding of these Orders is vital as the key to understanding Classical architecture. Summerson (1996: 12) states that: "The Orders came to be regarded as the very touchstone of architecture, as architectural instruments of the greatest possible subtlety, embodying all the ancient wisdom of mankind in the building art – almost, in fact, as

8.5   Illustration of Doric Order

products of nature herself." He also states that the Orders as used by the Romans were entirely central to their architecture, but were not used in a structural way. Instead they "make their buildings expressive, they make them speak; they conduct the building, with sense and ceremony and often with great elegance".

In contrast to Greek architecture where the temple was placed on high ground above towns and cities within a citadel or acropolis, Roman temples were invariably placed at street level and were very much part of the townscape. They were less ornate and usually did not have the crepidoma so typical of the Greek designs, relying more on a raised deep-side plinth and formal stepped frontal approach statement.

The Romans created new building types that had never existed before. They gave us the leisure delights of the spa and the bath seen at Bath and elsewhere in the United Kingdom. The Romans delighted in their leisure activities, but the culture of the bath was split into day and night activities. During the day the bathhouse was a meeting place set aside for business activity and debate. At night it became much more driven by pleasure than business. Summerson (1996: 37) suggests that these often complex structures were adapted by Victorian architects and became the "prototypes of public buildings of the railway age", such as the Pennsylvania Railroad Station in New York which imitated the central hall of the Baths of Caracalla.

The Romans also created an infrastructure system that survives today. Roads and evidence of surviving viaducts, aqueducts and bridges indicate that they were primarily engineers rather than artistic architects like the Greeks. They were masters of timber engineering and of defensive wall structures. Their experience with timber allowed them to develop the true or semi-circular arch by adopting timber arch

8.6   Illustration of Ionic Order

8.7   Illustration of Corinthian Order

formers or centering to temporarily support the arch until placement of its keystone permitted the arch to be self-supporting.

In addition to the extensive use of stone, the Romans also developed brick production. They produced bricks and tiles in vast quantities for their massive building operations, using them to create three types of wall construction (see Figure 8.11):

- Opus incertum
- Opus reticulatum
- Opus testaceum.

The Romans needed a constructional form and a material that allowed their armies, with the most basic of constructional skills, to build structures that underpinned their existence at locations far from Rome. That simple material was lime concrete, but the Romans developed its use to an extreme level. It facilitated the construction of major feats of engineering and structural form previously limited by stone technology. Buildings such as the Colosseum in Rome, the Pont du Gard viaduct at Nimes in France and, perhaps most famously, the Pantheon in Rome.

The Romans developed the use of concrete by accident. When building close to a town called Pozzuoli they used some of the local earth, which had sand-like qualities. They discovered that the resultant mortar was very much harder than anything they had previously used. The sand contained volcanic ash that incorporated elements of silicates and aluminates. These caused the mortar to set hydraulically as opposed to the normal carbonation process of pure limes. Roman concrete had certain limitations, notably it was poor in tensile strength but very strong in compression. However, the hydraulic qualities offered by these pozzolanic sands and aggregates provided a strong

8.8   Illustration of Composite Order

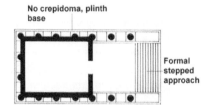

No crepidoma, plinth base

Formal stepped approach

8.9   Drawing of a plan of a typical Roman temple

building material. This development allowed them to extend their engineering skills and to construct buildings and structures that would otherwise have been limited in size and stature.

The advantages of this material are typified in the construction of the Pantheon. Its construction was facilitated by the use of concrete, even using lightweight aggregates at the upper levels of the dome to reduce its weight. At the time of its construction, it was the largest

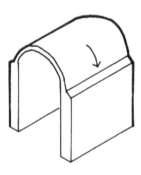

8.10   Typical Roman barrel vault or semi-circular arch

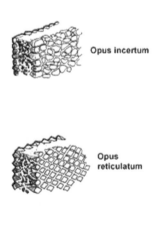

Opus incertum

Opus reticulatum

Opus testaceum

8.11   Drawing of types of Roman wall construction

8.12 Pont duGard. Nimes, France

spanning dome structure anywhere in the world. It still ranks amongst the five greatest domes in the Western world, sitting alongside Hagia Sophia in Istanbul, St Peter's in Rome, the Duomo in Florence and St Paul's in London. In design terms, the Pantheon may be compared with the Jefferson Memorial in Washington, this being another good example of Rome's enduring legacy in architecture.

The Romans were also the first civilisation to develop the use of concrete for arched construction. Indeed, one of their great achievements was their evolution of the semi-circular arch from its beginnings as a simple arch over an opening into a sophisticated form. Tunnel vaults, groin vaults and domes were all developed although they did not make the transition to the dome over a square plan form so typical of the subsequent Byzantine forms. The groin vault, however, was particularly important because it allowed the removal of the heavy walls needed to resist the forces in thrust generated by the semi-circular vault. In the groin vault the forces are directed to the four corners of the structure, allowing buildings to be lighter in structure and in terms of natural light into the building. This was achieved through the removal of the continuous flanking heavy support walls of the semi-circular vault, providing the forerunner of the gothic arch of our later medieval cathedrals.

The semi-circular arch was adopted as the fundamental construction form for the Colosseum in Rome, so permitting the construction of the auditorium terrace. This arrangement is familiar to us in our modern auditoria and is redolent of the amphitheatre in Greek culture. Construction began in AD 70, initiated by the Emperor Vespasian and was completed some 12 years later by Domitian. It has a huge elliptical plan form some 620 feet by 513 feet. The terrace seating is built up on a series of vaulted corridors, both radial and concentric on plan, adopting the use of the semi-circular arch referred to above. The constructional form allowed a capacity crowd of up to 50,000 to empty from the building in a matter of minutes.

The most famous of Roman architects was historian, polymath and student of Classical architecture Pollio Marcus Vitruvius (46–30 BC) whose book *De Architectura* is probably the most widely recognised and comprehensive treatise on Classical Greek and Roman architecture. He was hugely influential in the study of architectural influence and is a primary source for students and academics of architectural history. His written work (comprising the *Ten Books of Architecture*) has influenced a succession of later architects in their search for perfection of Classical architectural form, together with understanding its structure, influence and its relationship to landscape.

In Western Europe, the Romans colonised what we now know as Spain, Italy, Germany, France and England. The Romans combined both engineering and architecture to form their built environment and they adapted their skills to suit whatever materials were available. However, unlike the Greeks, they were more interested in expansion and impression than creating architecture for its own sake. This should by no means decry their architectural achievements and interest in the Arts, but it does help to understand the Roman culture by comparison with the Greek.

"The Romans came into the world with a sword in one hand and a spade in the other" (Seneca). The legacy of the Romans in Western Europe was their skill in town planning and the idea of the capital city. They also provided systems of law and order and administration. In Britain their reign lasted from about 43 BC to about AD 400. By AD 402 Roman soldiers started to leave Britain and by AD 410 Britain and Europe lost its Roman masters. On their decline they left a void that was to take a long time to fill. Their influence was undoubted and even in the nearest period of architectural style, the Romanesque of the tenth and eleventh centuries, their name and style (that of the semi-circular arch) perpetuated and formed the basis of Western European architecture which eventually developed into a Gothic form.

Eventually, and as a result of their long and geographically extensive period of dominance, the Romans both affected and absorbed cultural influences and were, in a reverse process, affected and changed by the peoples that they conquered. By the end of their dominance it was difficult to identify a true Roman as they had interacted in a truly symbiotic way and had, to some extent, become subsumed into the societies that they conquered and created.

It would be too easy to sum up Roman architecture as being defined by the semi-circular arch, the groin vault and the use of concrete. The influence of Roman architecture has been absorbed in many later periods of building development including the Renaissance, Palladianism and Neo-Classicism. Some of the exuberance of Rome is reflected in Baroque and Roman influence and styles have been subsumed in many, many later periods and by many, many later cultures. An understanding of Classical Rome is therefore central to an appreciation of architectural development.

8.13   Typical groin vault

Chapter **9**

Byzantine and early Christian architecture (third to eleventh centuries)

The Roman Emperor Constantine (306–337 BC) assisted the transition to early Christianity from Paganism when, in the Edict of Milan, he officially recognised the standing of the Christian Church. It subsequently became the official religion of the Roman Empire. Constantine only converted to Christianity at the very end of his life and moved his capital from Rome to Byzantium when under threat of attack in Rome in 334 BC. Byzantium was re-named Constantinople in Constantine's honour and is what we now know as Istanbul, Turkey.

The move from Rome to Byzantium allowed Eastern and Western cultures to merge, resulting in Byzantine architecture and art, and the creation of the Orthodox Church. In contrast, the migration of the Christian Church west and north from Rome resulted in the totally different architecture and art of the Gothic form, based on the plan of the Roman basilica.

The city of Constantinople, at the confluence of the two cultures of Roman and the Ancient East, developed rapidly. It remained fundamentally Christian from around the fourth century until it fell to the Turks in 1453. This Christian influence was expressed in the construction of numerous churches and religious buildings over the 1,000 years of its development and was the Christian centre of the then-known world. The artistic abilities of the Greeks were married to the engineering and building abilities of the Romans. It was not just the built form of the Christian Orthodox church that developed in Byzantium but also its culture, structure and teachings.

9.1   Hagia Sophia, Istanbul, Turkey

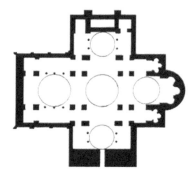

9.2   Typical plan form of the Greek Cross

The key to Byzantine architecture was the development of the dome. Its form extended beyond that of the circular plan form of the Roman dome into a plan based on the square. Probably the greatest example of this plan form transition is the magnificent church of Hagia Sophia in Istanbul (Constantinople), built in AD 537 by Emperor Justinian I.

Roman domes, limited as they were to a circular plan form, required substantial walls to take and distribute the massive loads and forces generated by the roof. Their limitation was that the need for such thick walls precluded the use of openings to provide light at low levels. The development of the dome to cover a square plan form was the great achievement of Byzantine architecture. The loads generated by the roof could be dispersed into four corner columns via over-sailing arches, supporting the dome, allowing an opening out of the lower levels of the structure and thus freeing up the plan form. With this development, square plan forms could link to square plan forms and allow linked unrestricted spaces, creating the Greek cross plan form that is so typical of the Byzantine church.

The key to all this was the 'pendentive'. The pendentive is the triangular structure that permits a square plan form to support a contained diameter dome. If four arches supported by columns in turn support a circle of the contained diameter of the square, small triangular sections are formed below the dome and above the arches at the corners.

9.3   St Mark's, Venice

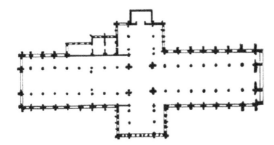

9.4   Cathedral plan form for comparison with Figure 9.2

The ability to form a support structure across these spaces – the pendentive – is the key to Byzantine square plan structures. A later, and the best-known example, of the Greek cross and domed plan form is that of St Mark's in Venice (started in 1063).

The stubby Greek cross-linked domes typical of the Byzantine church can be compared and contrasted with the elongated cross of the plan form of Gothic cathedrals such as Wells Cathedral. The plan form differences of the Western Christian cathedrals and the domes and Greek cross plan of the Byzantine church are also analogous of the schism that occurred between the Church of Rome and the Orthodox Church of the East. As Orthodox Christianity moved north into Central Eastern Europe the domical influence of the Eastern church architecture followed the migration. We see this in the domes over drums of the typical Middle European church so often represented in photographs of Romanian, Russian and Ukrainian churches and, eventually, stylising into the onion domes of the Kremlin and other iconic Eastern European buildings and churches.

In Western Europe the development of plan forms followed the basic concept of the Roman basilica, with long nave, crossing north and south transepts and chancel. The religious ceremonial also took on separate forms and adopted its liturgy to suit its plan form and vice-versa. The centralised and open ceremonials of the Byzantine Church, with its altar under the crossing, compared with the closed ceremonial of the Western Church, carried out in partial secrecy within the nave and behind the chancel or rood screen, dictates a different liturgy and approach to religious ceremony.

While externally Byzantine churches were plain, they had dramatic interiors of brilliant mosaic, often using blue and gold. These interiors depicted religious images and this art was developed to a great extent compared to other Christian locations. Even the Hagia Sophia in Constantinople (Istanbul) was heavily decorated by mosaic icon art. Istanbul was captured by the Turks in 1453 and the once-Christian church became a Moslem mosque with added minarets, as have become many similar, originally Orthodox churches in this part of Europe. Hagia Sophia church is now covered over by plasterwork and Islamic patterns and decorations as the original images would be unacceptable to Islam.

Chapter **10**

Early British medieval architecture

Having outlined the architectural influences from ancient civilisations to Byzantine architecture we can now turn to the styles of the United Kingdom. While it is tempting to categorise periods of architectural development by date, these should only be used as an indication of styles and not be used prescriptively. Architectural development should be regarded as a continuum and the dated periods given are therefore only indicative of generic dates when styles were fashionable, developing or in decline. Dates of various periods of architectural influence will also vary across the different geographical and cultural areas of Europe.

The medieval dates applicable to Britain are:

- Dark Ages/Saxon period: late sixth century to 1066
- Romanesque or Norman: 1066–1200
- Early English (Lancet or First Pointed): 1200–1300
- Decorated (Second Pointed): 1300–1400
- Perpendicular (Rectilinear or Third Pointed): 1400–1600, also sometimes called Tudor.

With the decline of the Roman Empire, much of Europe settled into a period of stagnation during which little significant architectural development occurred. This period is known as the Dark Ages. It was a time of warring and instability, particularly with Viking raids in the eighth and ninth centuries. In Britain, there was a period of church and secular development that we identify as Saxon; however, few buildings dating to this period survive. Examples include Escomb Church, County Durham (*c.*680) and Earl's Barton, Northamptonshire (*c.*1020–1050). Many of these Saxon buildings were constructed of timber and do not survive, and even the masonry examples are not numerous. They were typified by simple church buildings with small windows, sometimes with triangular-headed openings. Quoin

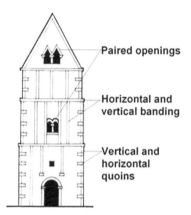

Paired openings

Horizontal and vertical banding

Vertical and horizontal quoins

10.1   Drawing of typical Saxon architecture

decoration was invariably in the form of long and short work with the quoins alternately being placed horizontally and vertically, producing a pattern towards the corners of the building. Window openings were typically paired, sometimes with triangular-topped openings. Curl (1999) states that an impressive surviving example is All Saints' Church at Brixworth in Northamptonshire which has recycled Roman bricks and tiles in the nave-arcades.

Chapter **11**

Romanesque/Norman in Britain (1066–1200)

The origins of Romanesque architecture are in France in the late eighth and early ninth century when a re-germination of culture and architecture began under the influence of the French King Charlemagne and the great bishops and abbots of the Christian Church. They funded what was probably the greatest period of religious architectural development in the Western world, from Romanesque to Gothic over a period of some 600 years. Influenced by the Roman basilica plan with its long vistas and side aisles, the buildings of this time began to show the fundamental format for the great cathedrals of the medieval period.

The Norman invasion of 1066 was a pivotal point in the development in England. William the Conqueror (1028–1087) implemented a survey in 1086, known as the Domesday Book, which gave the king a record of England, providing him with a fiscal overview of his conquered nation. He also introduced a feudal system which was essentially a pyramid of power with the monarch at its head and all others being subservient down to the simple serf and villein at its base. The king was considered to have divine precedence over all others and he handed out lands and granted ownership. Only nobles and the Church were able to hold land. This right by grant of the monarch remained until post-Magna Carta (a Royal Charter of political rights given to rebellious English barons by King John in 1215).

David I of Scotland (1124–1153), who was educated in the English court, also brought a version of the Norman feudal system to Scotland, with the king being the "feudal superior" over all lands, which he disposed of in return for payments and services from his nobles.

Under encouragement from the monarchy, there was a huge expansion of monastic buildings generated by the formation and proliferation of the monastic orders such as the Cluniacs, Benedictines,

Dominicans, Cistercians, Carthusians and Premonstratensians. All these religious orders owned vast tracts of land under the auspices of the king and within the feudal system. Enormous monasteries were established and a system of agriculture was developed to supply not only their own needs but also those of adjacent communities. They became landlords to tenant farmers and took tithes as payment for leasing their lands for agricultural use. These tithes or taxes were used to help to support the Church. They effectively replaced the Roman administration and provided a faith-based armature for society in support of an aggressive and dominant monarchy. The monasteries and religious orders were based on the Church of Rome – what we now identify as Catholic.

Religion was central to people's lives and the construction of numerous cathedrals during this period demonstrates this importance. As noted previously, the form and layout of these buildings was directly related to the nature of worship which tended to be more secretive in the Western Church than the hierarchical pattern of the Byzantine Church of the East. Furneaux Jordan (1997) asks "Did liturgy dictate architecture, or vice versa?" Here the clergy undertook religious ceremonies within the chancel, divorced from but overlooked by the laity within the nave. Even in the later nineteenth century, the upkeep of the two areas of the common church could be funded separately; the nave by the people of the church and the chancel by the local lord and the clergy in the form of the Diocese. In medieval times the naves of many churches were used for secular gatherings and local meetings – a much more adaptable use than their purely religious functions of today. Medieval naves had no seating and pews did not begin to appear until the late fifteenth century. In this context, it is interesting to consider how many of our great cathedrals do not have permanent fixed pews.

The ceremony of the early Christian church was processional, which is evident in their layout. The processional-facilitating plans of the aisles and the ambulatory surrounding the chancel was also extended to encompass the transepts in some cathedrals, such as at Winchester or Canterbury. The plan form of many of these great medieval cathedrals is clearly adapted from the Roman theme of the semi-circular and groin arch together with the basilica plan form. These cathedrals have a long central nave and associated side aisles, added to by transepts and crossing, a chancel with both apsidal and square end, and rotating chapels such as at Norwich, Gloucester and Westminster (although Westminster is later Gothic rather than earlier Norman).

Cathedrals also reflect different social history patterns. For example, English cathedrals were invariably constructed separately from their contiguous towns within a wall-enclosed space known as a "close". French cathedrals were part of the street scene, not at all separated and very much part of the town landscape. A comparison with the Greek acropolis temples might be drawn against the town location for temples chosen by the Romans. The choice of elevated temple sites in Greece might be likened to the separation of religious and secular life in England, whereas the 'street' temples of Rome might be likened to the location for French cathedrals in the hub and mêlée of the town.

Turning to architectural styles, the earliest Romanesque churches, particularly those constructed during the eleventh century, were massive and plain with little articulation and decoration of their struc- ture. However, as time passed the masons and the sponsors of church and cathedral buildings developed a new or improved style that added character and decoration. Large circular columns and flat support columns morphed into a more delicate appearance and less massive structure. Certainly after 1100 there was an increase in the level of decoration adopted.

This use of decoration had a practical purpose in addition to being aesthetically pleasing. The level of literacy in the mass of the common people was very low: few could read and even fewer could write. The stories of the Bible therefore had to be transmitted in an 'acceptable' form to the laity. This was achieved by making religious buildings provide the story in pictorial form: the stained glass windows and carvings provided a canvas to transmit biblical stories and parables. Decoration also takes the form of Classical-inspired motifs and even Celtic animals and mouldings indicating the variety of influences during this period.

These early Romanesque churches and cathedrals had very simple arches and columns, square or round in plan with simple, square cushion capitals. Usually they were two-storey structures, three storeys becoming usual in Britain, with a clerestory above the ground- level rows of columns and walls. It was not until later that the third tier was added as a gallery floor (the triforium) above the aisles, with a clerestory above that provided light at high level, extending height and producing vertical emphasis that is so typical of our later Gothic cathedrals.

The use of the semi-circular arch is exemplified at Norwich Cathedral. This is one of the most complete Romanesque buildings in Europe, the foundation stone having been laid in 1096 and the building com- pleted in its early form by 1145. The west front at Norwich Cathedral demonstrates typical Norman semi-circular arches with carved deco- ration on the secondary doors left and right of the main door. The main west window in the centre is more typical of later Decorated style. Different coloured stone is evident in the main west door archi- trave, indicating a later alteration, as is the main west front window. The darker stone is probably from Nottinghamshire (Barnack), whereas the original cathedral was constructed in paler Caen stone.

11.1   The north wall at Norwich Cathedral (Norfolk, England) showing typical Norman semi-circular arch
form

The pilaster ribbing was also probably added later, indicating the great complexity of these historic buildings and how difficult they can be to interpret.

The semi-circular and groin arch has particular limitations in that it requires thick heavy walls to support the roof and the thrust from the semi-circular arch and groins. It therefore produces a dark and dingy space limited by the need for the supporting walls to be so thick and heavy.

Cathedrals may demonstrate a variety of architectural influences, particularly because they were constructed over such long periods. What may have originated as Romanesque or Norman may be influenced by later fashions and styles so that, as the building increases in height and over the time taken to construct, each level may demonstrate a different architectural influence. Inevitably, because these great cathedrals took many decades or even hundreds of years to complete, the resultant eclectic stylistic influences reflect the numerous influences and changes that occurred over their period of construction. For example, construction on Durham Cathedral began at the end of the eleventh century but it took many decades for

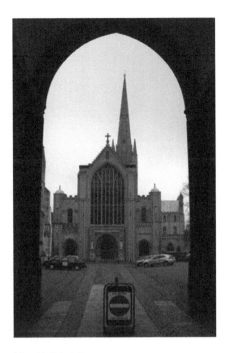

11.2   The west front of Norwich Cathedral

11.3   Norwich Cathedral aisle groin arches

11.4   Norwich Cathedral, choir and chancel

11.5   Exeter Cathedral, England

it to be completed. The Galilee Chapel was begun almost 100 years after the original foundations were laid and so exhibit more decorative stonework, such as dog-tooth carving to the arches which are indicative of later Norman decoration. The central tower of Durham Cathedral was rebuilt in 1490 following the destruction of the original tower by lightning and major works were undertaken in the late eighteenth century due to cracks in the vaulting. Sometimes work was therefore as a result of the need for urgent repairs and these would offer the opportunity to enhance or alter the original design, giving these buildings numerous layers of architectural influence. At Exeter Cathedral, the two towers are very clearly Romanesque in design and date between 1112 and 1200. However, the rest of the cathedral is Gothic, showing Early Decorated (1270–1328) and Late Decorated (1328–1375) architectural influences and thus demonstrating the long period of building involved at this site and the associated changes in design influences.

In other locations dramatic events sometimes meant that cathedrals had to be reconstructed, altering their architectural style. Lincoln Cathedral was consecrated in 1092 but suffered a dramatic fire in the early twelfth century, followed by an earthquake in 1185. Such events mean a reappraisal of styles and materials. The resulting building can be extremely complex.

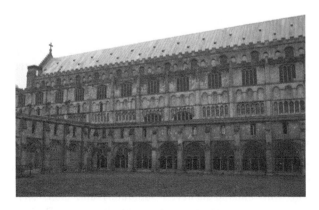

11.6   Norwich Cathedral south wall from cloisters. Many periods of Gothic styles are demonstrated as well as the original Romanesque

11.7   Lincoln Cathedral, England

The smooth-sided columns and undersides of the early Norman arch became articulated and 'lightened'. This development meant that timber "centering", or the temporary timber support framing necessary to hold up the arch until the keystone was completed, could become much lighter and more portable – there was no need to provide a temporary timber support frame over the whole of the inside of the structure. Centering could be used and moved as the structure was advanced. The whole process of building therefore became more efficient in terms of both time and cost.

The next development was the adoption of ribbing at the junction of the groins. This permitted only the groin joints to be supported by centering, with infilling achieved separately and later. The earliest example of high rib vaulting in Europe may be seen at Durham Cathedral. Here the semi-circular arches are taken diagonally across the square plan of four columns, helping to distribute the loads generated by the roof more effectively. At each outer pair of columns the resultant lateral arch is pointed because of the shorter span distance across the nave between columns. The outer walls of the cathedral also demonstrate the first use of flying buttresses. Impressive buttresses are also evident at Dunfermline Abbey. Created by King Malcolm Canmore (1057–1193), the Norman nave was built using stonemasons from Durham.

In Scotland early ecclesiastical architecture demonstrates the influence of the Celtic Church from Ireland, where the monastic movement was centred around the island of Iona. St Columba arrived in 563, the current cruciform church dating to the early 1200s. The Irish influence is evident in the early round towers dating to the early eleventh century found at Brechin, Angus and Abernethy, near Perth. The tower at Brechin was then incorporated as part of a medieval cathedral, but at Abernethy the tower stands as it was intended, alone. Such towers were places of defence for church treasures. A high entry-door was probably accessed by a movable ladder. The church building, probably a simple timber structure, would have stood separately. Similar towers are found in Perthshire and Fife at Muthill, Dunblane, Dunning, Markinch and St Andrews (St Rules) but are square in plan with twin light belfries and date to the early twelfth century. The medieval

11.8   Durham Cathedral

cathedral at Dunblane and the now ruinous church at Muthill both sit awkwardly against their later church buildings, indicating that they were originally solitary towers. Culdee monks reputedly were associated with these churches and particularly at Muthill where this religious order existed to a much later date than in other parts of Scotland.

These early square and round towers are indicative of the influences on Scotland's developing early churches and associated buildings. The round towers hail from Ireland and were then adapted into a more Romanesque square tower and subsequently incorporated into medieval churches (with the exception of Dunning where the tower was probably built with the church). This demonstrates the geographical variation in architecture which was a continuing theme. In many respects Scotland had stronger associations with the Continent than England and this is evident in the architecture. Fenwick (1974: 23) states:

> Scots masons never quite gave up their affection for the rounded as opposed to the pointed arch. It crops up again and again throughout the Gothic period, and returned in full flavour at the Renaissance. This fondness may have reflected a harking back to Celtic modes, they having Italian, or at any rate, Roman precedent.

11.9   A round tower, Abernethy, Scotland

11.10   St Rules at St Andrews, Scotland

11.11   Dunning Parish Church, Perthshire, Scotland

11.12   East Anglian round tower church at Fishley, Norfolk, England

It is conceivable that the small number of Scottish round towers surviving may have been influenced by Irish tower builders. In England, however, there are considerable numbers, mostly in East Anglia, but the debate is on-going as to their origins. Whether they were constructed originally as defensive towers against invaders from the Continent or as bell towers or whether they were forced as a constructional form because of the absence of endemic stone in East Anglia, remains a mystery.

During the medieval period it is estimated there were about 1,000 round towers in the United Kingdom (Goode, 1994). Today there are approximately 125 round tower churches in Norfolk, 47 in Suffolk and only 13 in the rest of the United Kingdom (Morley, n.d.). By comparison, Goode documented 181 round towers in the United Kingdom during the 1960s and 1970s, with a total of 175 in East Anglia. This tends towards the argument that their concentration in East Anglia might be explained because of the absence of stone to construct stable corners. Flint is used as an infill material. Flint is a commonly used material in Norfolk but it has its limitations in forming structural corners, hence the adoption of a round tower plan form. Speculation has it that until the Norman invasion in 1066 the preceding Saxon people of East Anglia did not have the transportation capability to import stone and, via its use, construct stable square corners to their buildings.

The debate is also on-going as to the original use of round towers. What does seem odd is that churches attached to these round towers do exhibit the use of stone used in construction of corners of both naves and chancels. Perhaps this adds weight to the argument that the towers pre-date the churches and that the churches made use of existing round towers to provide their bell towers. The usage of these possibly pre-existing towers remains an unanswered question.

Throughout this early church development, the role of the mason became paramount and the union of the craft skills became closely allied with the requirements of the church. It was during the thirteenth and fourteenth centuries that the crafts started to form into guilds and associations to further their interest and maintain a hierarchy of abilities and levels of skills – the apprentice, the journeyman and the master craftsman. The Association of Freemasons was founded in the early thirteenth century, demonstrating their pre-eminent position in early medieval society. These craft guilds were central to the medieval social structure and their ceremony and structure can be seen in some of our contemporary associations, such as the Freemasons.

Chapter **12**

Gothic architecture 1200–1500

Norman architecture may be regarded as heavy and dark with massive walls and columns resisting thrust from the semi-circular form of the roof and vault. This imposed the adoption of a square plan. However, in Gothic architecture the use of a pointed arch of varying curvature allowed the creation of a much more open structure, full of light. Here the structural forces are in equilibrium by transference from one element to another – from the web of ceiling vaulting, down columns and thence dispersed to aisle and ground via buttresses and flying buttresses weighted by pinnacles and towers, as exemplified by Exeter Cathedral. Windows are wide, only limited by the rhythm of the piers. The square plan form of Norman churches was therefore liberated to become the open rectangular plan of the Gothic through the development of the pointed arch. This allowed the cathedral builders of the early thirteenth century to develop an unfettered celebration of form, to bring light into the contained space and introduce new height without the mass and weight that had previously been a structural necessity. If Romanesque was austere then Gothic was exuberant.

Gothic cathedrals adopted a vaulted roof with steep arches in section across the shorter side and flatter arches across the long spanning side; all terminating at the same ridge height by variation of the curve of the arch.

In English architecture there are four periods of Gothic form (Curl, 1999):

- Early English or First Pointed: from the end of the twelfth century to the end of the thirteenth century lancet windows were notably used, later incorporating some plate tracery.
- Decorated or Second Pointed during the fourteenth century featured geometrical tracery: curvilinear, flowing or reticulated. In England this period was short-lived.

12.1   Flying buttresses and pinnacles, Exeter Cathedral, England

- Perpendicular overlapped with Decorated and was present from *c.*1332.
- Tudor during the mid-to-late sixteenth century.

Chartres Cathedral in France is one of the greatest Gothic cathedrals. Although the towers and Royal Portal are twelfth century and exhibit Romanesque decoration, the majority of the cathedral dates to the thirteenth century following a major fire of 1194 which had largely destroyed the Romanesque cathedral. The cathedral is famous for its outstanding stained glass including the thirteenth-century rose windows. Lancet windows situated above the Royal Portal demonstrate a transition from Romanesque to Gothic but much of the cathedral demonstrates its Gothic roots, having been rebuilt after the fire in less than 30 years. It should be noted that architecture in Europe developed slightly differently from England at this time and, therefore, the styles do not necessarily correspond in the same way.

At Exeter Cathedral, the Gothic style dating 1270–1369 is demonstrated in the vaulting of the ceiling. Here the impressive design is recognised as the longest unbroken section of Gothic vault to be found in the world. Buildings like Chartres, Durham and Exeter cathedrals exemplify the fact that the craft of the stonemason was developing almost into an art form, with ever-increasing delicacy of details more applicable to wood carving than unyielding stone. This may facilitate a parallel to be drawn with the Ancient Greek masons whose abilities were more dictated by stone as an art medium than a structural material. This ability reached its zenith during the sixteenth century in the gossamer-like web tracery of the Perpendicular fan vaulting of that period. This is evident in the Henry VII chapel at Westminster.

We can see when considering the development from the earliest Gothic form to the last and most exuberant phase of it – Perpendicular/ Tudor – that there is a transition from a religion-dominated English society to a secular-focused one terminating in the Dissolution of the Monasteries under Henry VIII in 1538–1541. At this time, religious authority was transferred from the Church of Rome to the Monarch and it represented probably the greatest period of religious upheaval that England has ever seen. Many monasteries and abbeys were totally destroyed or left to fall into ruins. The period is known as the Reformation – the reform of the religious focus and culture in

12.2   Fan vaulting

12.3   Gothic vaulting, Exeter

12.4   Ruins of Thetford Priory, Norfolk, England

England. The period spans four monarchs (Henry VIII, Edward VI, Mary and Elizabeth I) and witnessed, under Edward VI the reinforcement of the English Protestant Church via the publication of two prayer books. The Irish Church remained Catholic.

In Scotland the Reformation was later than in England, James V having initially resisted the move to follow England's lead. However, in 1559 in St John's Kirk in Perth, John Knox delivered a sermon against idolatry resulting in angry mobs attacking the local monasteries, although St John's itself was largely spared. Although there was some destruction of religious buildings, the loss of many of Scotland's ecclesiastical buildings was not as a result of the Reformation and the move to a Protestant Church. Instead it was partly due to lack of investment in the buildings which resulted in their continued decline. The nave of Dunblane Cathedral was roofless for almost 200 years before being restored in the late nineteenth century, but this was due to a continued lack of investment and is not attributed to the work of rioting mobs. Other important ecclesiastical centres like Arbroath suffered the same fate.

However, the impact on the design of church buildings was certain. Fawcett (1994: 334) indicates that there was a lack of money for new churches but that those which remained were often too numerous and not suited to the new form of Protestant worship. One of the first buildings to be purpose-built for the new style of worship was at Burntisland, Fife in 1592 and was an "innovative centralised building with simple Classical detailing" and which "pointed the way to a radically new approach to planning" although many that followed were not as "revolutionary" in their plan.

12.5   Dunblane Cathedral, Dunblane, Scotland

12.6   Ruins of Arbroath Abbey, Arbroath, Scotland

Chapter **13**

Early British domestic architecture

While all this religious development was taking place, our early secular society was also in transition. The medieval world was a dark and aggressive place with constant feuding between warring factions. Its secular architecture reflected this pattern of continual feuding with the building of castles and fortified structures, their main focus being defence against attack by enemies. However, from the fifteenth century onwards the focus of secular architecture on defensive constructions – castles and fortified houses – subtly morphed into development less centred on defence and more on comfort and quality of settled domestic life.

During the medieval period we also see the beginnings of and focus on the three areas of British social structure: the English, the Welsh and the Scots. Although the Romans under Hadrian moved into Scotland (Scotia), they took a defensive line (Hadrian's Wall) in the north of England in what we now know as Cumbria. Similarly, Offa's Dyke offered a defensive barrier against the Welsh. Ireland remained a distant and isolated land beyond the stretch of water we now identify as the Irish Sea. Within England, with the departure of the Romans in about AD 410, feuding between neighbours in pursuit of land ownership and power continued. So, it is hardly surprising that domestic secular architecture had its focus on defence. Towns and large settlements were constructed within defensive walls and many of these remain as examples today, such as in Norwich and Exeter.

Immediately following the Conquest of 1066, it became necessary for the invading Normans to construct positions of defence against marauding attacks. Early defensive structures were provided by timber-built, motte-and-bailey fortifications. The bailey was a yard or enclosed space with an outer defensive wooden wall of stakes, sometimes with a ditch or moat in front, surrounding an earth mound or motte surmounted by a defensive keep or tower. A similar

13.1   Part of the flint and brick defensive walls of Norwich, England

13.2   Flintwork-constructed round tower (Cow Tower, thirteenth century) forming part of the Medieval defensive structures of Norwich

arrangement is evident at Durham Castle. Here a defensive mound was initiated by William the Conqueror in 1072 but was occupied by the Bishops of Durham for many centuries. The former Norman stronghold is now occupied as part of the University of Durham, having been renovated to form student accommodation.

These early defensive structures provided the site and genesis of the stone-built castles that we are more familiar with, although many of the latter were re-sited on rocky outcrops or adjacent to a river; both providing for more easily defended sites. It was during the thirteenth century that the greatest period of castle construction occurred, although important sites such as the Tower of London showed earlier development of permanent structures, including a timber defensive structure commenced in 1066. The first stone construction, now known as the White Tower, was completed in 1097. These stone structures adopted a concentric plan with outer walls and elevated walkways surrounding a bailey with an easily defended inner keep or tower.

Dover Castle in Kent, built in 1180, is a typically heavily defended structure with massive protective walls and a substantial castle within. Other heavily defended structures include Beaumaris and Caernarfon in Wales.

Scotland is particularly well known for its large number of surviving castles and defensive tower houses. Following the defeat of Edward II at Bannockburn, Scotland turned to France as her main ally, resulting in a much more continental style with turrets, conical roofs and decoration reminiscent of French chateaux. This style was then revived by the Victorians in the Scots Baronial style where they revisited these medieval castles and used the turrets, conical roofs and dormer windows to evoke a link with the past, albeit not a historically accurate one. However, the constant feuding with England and locally within Scotland between landowners and clans, meant that defensive structures prevailed but gradually they became more domestic in style, as homes rather than keeps. The move from the fourteenth-century defensive keep to a much more comfortable family home at Drummond Castle, Perthshire is indicative of this transition to a more settled society.

While town walls were also indicative of the need for defence, these became important for controlling access in and out of towns through the town ports or gates. The walls were not required to prevent attack by an enemy, but instead allowed taxes and tolls to be collected by market traders entering the burgh. Town ports could also be closed during times of disease such as plague.

Like cathedrals, these buildings demonstrate many different stylistic influences and periods of development and decline depending on the wider economic and social situation at the time. A castle like Stirling, situated on a craggy defensive rock in the heart of Scotland, shows many layers of development and change over time. The defensive and strong construction of these buildings, however, means that they do survive today, albeit in a much altered form from their original construction.

Although the built evidence of fortified architecture is still visible today, the domestic architecture of the common people would have been very simple thatch, wattle and daub, and timber structures which do not survive to the present day. Later medieval timber-framed houses do survive in parts of England although few remain in Scotland. These will be examined in a later section on vernacular buildings.

It is clear that the great medieval period of religious architectural development was brought to an abrupt and destructive end during the Reformation. It is at this time that we also see a change from religion-focused building development to more secular development. The monarchy remained powerful and the nobility were sufficiently rich and influential to commission large houses and mansions without the need for their primary function to be centred on defence. The whole focus of society was in transition from combative and religious architecture to one with a greater focus on society in general.

13.3   Norwich Castle and its defensive man-made mound or motte

With a more settled society, buildings associated with trade were erected in towns, some of which were funded by merchant and guild interests. The craft guilds and merchants became important elements

13.4   Durham Castle, Durham   **13.5**   Drummond Keep with the house on the right, Perthshire, Scotland

13.6 A town gate in York, England

13.7 Stirling Castle, Stirling, Scotland

in society during the fourteenth and fifteenth centuries; the former as a result of the extensive church development period between the eleventh and early fifteenth centuries and the latter with an increase in trading. They had a powerful status within society and, as a result, numerous important buildings were constructed under the guilds' and merchants' direction during the fourteenth and fifteenth centuries. The increasing involvement in administration of towns and cities meant that buildings such as guild halls were erected.

13.8 Norwich Guild Hall, Norfolk, England

Chapter **14**

Conclusion

This part has outlined what amounts to several thousand years of architectural development. The architectural styles inspired by the ancient civilisations of Egypt, Rome and Greece have, at various times, waxed and waned in popularity. However, they undoubtedly still have a powerful influence on architecture today. Revered and respected, their architectural legacy is worthy of more detailed investigations by all students of architecture and the reader is encouraged to seek more detailed texts on this topic listed in the Bibliography. An understanding of the Classical Orders, their origins and development is relevant for many later architectural periods and for a variety of building types.

In Britain, the Romanesque style was adopted in the eleventh century, but by 1200 Gothic emerged and dominated English religious architecture for 300 years. In contrast, domestic architecture was dominated by heavily defended structures built in prominent defensive positions. These reflected the unstable society and feuding factions which dominated early medieval life. However, gradually, with increased stability, the guilds and merchants emerged to dominate architecture in towns as the power of the Church receded in favour of trade and commerce.

So, constantly intertwined in the development of our heritage there is subsumed a cultural current that is as important to understand as the buildings that were produced. Architecture is a product of what is happening socially and culturally, and without understanding of one there cannot be erudition of the other. Our built environment is only a reflection of our society (much as literature and art) so what we have is a tapestry of a multitude of many-coloured threads. Only when this is stitched together as cloth do we understand the finished or developing picture. It is this cultural symbiosis and our understanding of it that underpins the need for a conservation response to our heritage and its on-going management in order to perpetuate it for future generations as well as giving respect and recognition to the sequence of events and influences that have formed it. Without an understanding conservationist response our heritage might otherwise be lost or its record damaged irretrievably.

Part 3

The Renaissance to the twentieth century

Part **3**

Introduction

While the ancient civilisations we examined in Part 2 were the foundations and influenced later architectural styles, it has really been in the last 600 years that these design influences were taken and used in the design of the buildings which are evident in our towns and cities today. This part will examine these architectural developments from the Renaissance to the present day.

Gothic architecture saw its germination in France during the late twelfth century and spread as an architectural influence across northern and central Europe and into Britain, where it influenced architectural style until the mid-sixteenth century. It was later revived as Neo-Gothic or Gothic Revival during the nineteenth century.

From the early seventeenth century Britain was also heavily influenced by the Italian Renaissance. Italy is recognised as a seat of tremendous architectural innovation and this is evident centuries later in the buildings which survive in the great Italian cities of Venice, Florence and Rome. Italian architecture was subsequently enormously influential throughout the rest of Europe where it was adopted and adapted to varying degrees. In Britain the Classical influence is evident in the buildings of the Georgian period.

Although these styles dominated architecture for certain periods, their influence on fashionable building design waxed and waned over time and sometimes resurged under periods of "revival". This part will identify these many influences and their complex patterns of change through time until the period following the Second World War.

Chapter **15**

The Italian Renaissance

From around AD 600 to AD 1400 Italy was dominated by Romanesque architecture. This was a style developed into the thirteenth century as exemplified by the complex of buildings at Pisa in northern Italy. These include the cathedral at Pisa, dating to the late eleventh century, the Baptistry built 1153–1278 and the famous Campanile bell tower or Leaning Tower built 1174. The Baptistry was designed by Dioti Salui and is on a circular plan and although there are later fourteenth-century Gothic additions, it is still Romanesque in design (Fletcher, 1928).

Italy's architectural influence remained founded in the Classical architecture of Greece and Rome, Romanesque styles used even until the Middle Ages when Gothic was flourishing in other parts of Europe. As a result, examples of Italian Gothic architecture are less extensive than in other parts of Europe but, over the period 1200–1450, there were some magnificent buildings erected in this style including Doge's Palace in Venice (1309–1424), Milan Cathedral (1385–1485) and Siena Cathedral (1245–1380). However, Banister Fletcher (1928: 499) notes that "the influence of Roman tradition remained so strong that the conspicuous verticality of Northern Gothic is generally neutralised in Italy by horizontal cornices and string courses".

From the fifteenth century, Italy experienced what is now termed the Renaissance, a cultural revival or rebirth of thinking and philosophy. The term is derived from the Latin languages of French – *renaitre* meaning to be born again or Italian – *rinascimento* meaning rebirth (Curl, 1999). The Renaissance saw its germination as a humanistic ideology and philosophy which subsequently developed into art, architecture, culture and learning. This was a period of enlightenment in a society with its roots still very much in Classical Rome as Furneaux Jordan (1997) describes: "The Renaissance was a great awakening and a great enlightenment. It was born in Italy because Italy had

15.1   The Leaning Tower of Pisa

known so little of the glories of Gothic ... and remembered ... the glories of the Roman Empire."

The Renaissance movement was sponsored by the increasing wealth of the numerous merchants and trading families based in the towns and ports of northern Italy. They were beginning to expand their trade routes from ports like Genoa, Pisa and Venice across Europe and beyond into the East and Arabia. Cities like Florence and Siena became focused as centres for trade, allowing merchant families to amass huge wealth. The Medici family (thirteenth to seventeenth century) were highly influential bankers and their support of Renaissance architecture created many important buildings of the period. Furneaux Jordan (1997) states that "they were the first great merchant patrons of the age". Other great mercantile families included the Pitti, Strotzzi and Pandolfini. They all contributed to the ascendance of urban Renaissance architecture by building large family palaces or Palazzo in the towns and cities of Florence, Venice, Genoa and Rome. In the countryside substantial villas were constructed.

The Renaissance sparked an interest in aesthetics and design based on Classical principles and seeded the role of the artist architect, laying down the foundation for the profession that we recognise today. Some of the greatest names in architectural history gained reputation, kudos and celebrity. These include Brunelleschi (1377–1446),

15.2   The Baptistry, Pisa

15.3   Doge's Palace, Venice

Alberti (1404–1472), Leonardo da Vinci (1452–1519), Michelangelo (1475–1564) and Palladio, born 1508 in Padua, followed by Bernini (1598–1680), who was closely identified with the Baroque school of art and architecture.

Filippo Brunelleschi is described by Curl (1999) as "the first and perhaps the most distinguished" Renaissance architect. Although his original career was as a goldsmith and sculptor, he won the competition to design a dome and cupola for Florence's cathedral, the Santa Maria del Fiore. His plan for the dome, known as The Duomo, was a remarkable feat of design and engineering, constructed in brick with an octagonal double shell construction. Brunelleschi also designed the Foundling Hospital in Florence (1419–1444) which Curl (1999) describes as "the first truly Renaissance building, but its sources are local". The building is believed to be the first hospital in the world to be built for foundlings and was originated by the Silk Guild, an example of one of the philanthropic duties undertaken by these guilds.

Renaissance architecture in Italy was highly influenced by Vitruvius (*c.*80–25 BC), the Classical Roman architect and philosopher. He defined the five essential design principles for Classical architecture as: order, eurythmy (beauty, harmony, rhythm), symmetry, propriety and economy. He advised on the need for *commodity*, *firmness* and

15.4   The Duomo, Florence

delight when designing a building and planning a site – especially in relation to buildings' aspect and interaction with landscape – the Vitruvian ideal. He defined beauty (*venustas*) via an association with the proportions of the human body. The famous *Vitruvian Man* of Leonardo da Vinci fame was originally analysed and drawn by Vitruvius. One of his most important works, *De Architectura*, written *c*.27 BC, was rediscovered during the fifteenth century and significantly influenced Italian Classical architects of the time.

Leone Battista Alberti (1404–1472) published his major work *De Re Aedificatoria* (*On the Art of Building*) covering the art and science of architecture, town planning and landscape design. It was first printed in Florence in 1486. Though not widely available outside Italy until the sixteenth century it became a primary source for Renaissance design philosophy. This work effectively repeated but extended that of Vitruvius and covered the same subject areas but, perhaps, in analysis and explanation rather than simply going over the same ground. Alberti also covered the philosophy and interpretation of beauty, as did Vitruvius. His volumes became the principle source for inspiration and study of art and architecture during the early Renaissance.

Born in Padua in 1508, Andrea di Pietro della Gondola, who was nicknamed Palladio by his mentor Gian Georgio Trissino, is, perhaps, the most influential Classical architect of the late Italian Renaissance. His treaties, *I Quattro Libri dell' Archittectura* (*The Four Books of Architecture*) were published in 1570. This work defined his philosophies of

15.5   Uffizi, typical Renaissance styling in Florence

architecture, influenced by both Vitruvius and Alberti. Palladio was influential in British architecture and his design philosophy is encapsulated in early eighteenth-century British Classical architecture. Summerson (1996: 136) says he is recognised as "the greatest modern interpreter of classical architecture". His buildings include many substantial villas including La Rotonda or Villa Capra at Vicenza (c.1566–1570) and Villa Barbaro at Maser in 1560. He also designed numerous churches including San Giorgio Maggiore (1564–1580) and Il Redentore (1576–1580), both in Venice.

In addition to the interest in architecture, the Renaissance also witnessed a great revival in learning, stimulated by reading and scholarly pursuits. The printing press was invented by Johann Gutenberg in the 1450s, which meant that the classics were available in print form and the Gutenberg Bible was published in 1455. The greater availability of literature facilitated by the invention of the printing press was, undoubtedly, a major factor in the spread of Renaissance culture, art and architectural influence in Europe during the sixteenth century. However, the Italian merchants were also instrumental in disseminating Renaissance styling as they pursued their financial interests across Europe where the style was widely adopted but also adapted to local styles, tastes and materials. In England, its adoption was set against a canvas of 300 years of well-established Gothic influence and was slow to manifest.

So yet again we can recognise the composite of cultural cause, effect and response reflected throughout history, determining our historic built environment. The need to fully understand this pattern of influence and how it subsumed art and architecture must be clearly understood if we are to value and appreciate the historic record and narrative that our historic built environment offers when properly recognised and translated.

Chapter 16

Renaissance architecture and the return to Classicism in Britain

Renaissance influence came late to Britain. With 300 years of continually evolving Gothic forms, from Early English through Decorated to Perpendicular, Britain relinquished its Gothic heritage with no little resistance. However, by the mid-sixteenth century the impact of Renaissance thought was gradually beginning to emerge and we start to see the germination of Classical influence in art and architecture. This then expanded during the next 200–300 years, with periods recognised as Palladian, Baroque, Neo-Classical, Greek Revival and Picturesque movements in Britain extending well into the nineteenth century. During this complex period of architectural history these styles evolved, merged, overlapped and sometimes re-emerged as revival periods. It is therefore perhaps helpful at this point to give a brief outline of the different styles.

Palladian is a Classical architectural style which was instigated by Inigo Jones in the early seventeenth century, although a Second Revival occurred in the early eighteenth century, encouraged by Lord Burlington and Colen Campbell.

Baroque is a very exuberant Renaissance-inspired style which used curving lines and convex and concave patterns. It was found in the seventeenth and eighteenth centuries and evolved into Rococo. Exponents in Britain were typified by Vanburgh, Hawksmoor and Wren.

Rococo is an exuberant style based on marine and shell motifs which emerged in eighteenth-century France. It was heavily influenced by the sculptor and artist Giovanni Lorenzo Bernini (1598–1680).

Greek Revival is a Neo-Classical style dating from the 1750s which was inspired by the study of Greek temples and other Ancient Greek buildings. The main architects were Adam, Playfair, Hamilton and Thomson. It became a particularly popular style in Scotland during the nineteenth century.

The Picturesque Movement was an aesthetic movement which emerged during the eighteenth century. It focused on landscapes and nature and moved away from the symmetrical focus of the Classical styles. Architects included John Nash.

Neo-Classical emerged in the late eighteenth century and extended into the early nineteenth century. It was inspired by archaeological excavations and the study of ancient buildings and sites, such as at Pompeii, and may be linked to Greek Revival.

The context and development of these different styles will now be examined.

Following the Reformation, England experienced a resurgence of wealth and influence under the reigning monarchs, Henry VIII and Elizabeth I. This is known as the Tudor period (1485–1603). The nobility and aristocratic members of society designed some of our most well-recognised and iconic great houses including Hardwick Hall, Blickling Hall and Hampton Court. Others were re-modelled and re-styled former family seats such as Wilton House and Grimthorpe Castle. Under the reign of Elizabeth I, and subsequently through James I (the Stuart period 1603–1714), England prospered. Elizabeth, as a strong monarch, did much to stabilise England, encouraging foreign trade, commerce and prosperity. The great houses constructed in this period were a celebration of that prosperity.

Examples of English buildings demonstrating some Italian Renaissance styling do not appear until the late sixteenth century. The Elizabethan surveyor, Robert Smythson's experimental Hardwick Hall, Derbyshire (1590–1597) is a masterpiece of glass and stone leading to the phrase "Hardwick Hall, more glass than wall". Smythson had previously worked as a surveyor at Longleat and was appointed by Bess of Hardwick to undertake the design of a new hall. At that time the term 'surveyor' (*surveyour* [sic]) was the more common definition of what we now understand to be the role of the architect. The style owes more to Elizabethan Gothic without the arch rather than solely to Italian Renaissance influence and is redolent of the adoption of the Gothic introduction of light through tall glazed windows. It makes a certain statement of horizontal emphasis in its use of projecting stone string coursing which breaks up the vertical lines of pure Gothic architecture. It might be loosely defined as a marriage of Gothic and Renaissance design and does mark a significant turning point in English architecture with a new leaning towards Italian Renaissance influence.

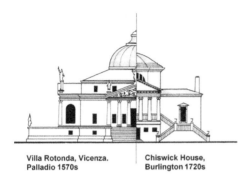

Villa Rotonda, Vicenza.   Chiswick House,
Palladio 1570s   Burlington 1720s

16.1   Chiswick House, Lord Burlington/Villa Rotonda

For a real British interpretation of Renaissance styling we need to look at Queen's House in Greenwich (1616–1635) and The Banqueting House in Whitehall (1619–1622), both by Inigo Jones (1573–1652). He was the Surveyor of the Kings Works in 1615 and is recognised as being the first important English architect. Inigo Jones was largely responsible for introducing the Palladian style to Britain as Curl (1999) explains: "only in the early seventeenth century was uncorrupted Renaissance architecture firmly based on Italian prototypes, introduced in England by Inigo Jones". The Banqueting House was subsequently re-fronted by John Soane in 1829 but retains the essence of the original 1619 Renaissance-influenced design of Inigo Jones. These buildings had an intense design relationship with their grounds following the inspiration of Vitruvius, Alberti and Palladio. This is exemplified by the Villa Rotonda (1550–1551) at Vicenza,

16.3   Queen's House, Greenwich

16.2   Holkham Hall, Norfolk, Palladian styling

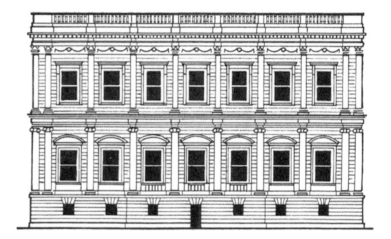

16.4   The Banqueting House, Whitehall – showing the Soane re-fronting of 1829

designed by Palladio and interestingly copied in close replica by Lord Burlington at Chiswick House, London in 1725. For pure Palladian interpretation we may look at Holkham Hall in Norfolk, built for the Earl of Leicester to William Kent's design, begun in 1734. Very early Renaissance influence might have been demonstrated by Henry VIII's designs for Nonesuch Palace at Richmond and by Old Somerset House in London (now demolished).

Chapter **17**

Scotland's royal palaces

In Scotland, the Renaissance was evident in the numerous royal palaces which were built primarily under the influence of the Royal Stewarts, James IV and James V. Fawcett (1994: 301) states:

> So far as secular architecture is concerned, the Middle Ages passed into the age of the Renaissance on a triumphant note with the construction of a series of royal residences ... and in them we find a swaggering architectural expression of the claims of the Scottish monarchy to a role on the wider European stage.

These buildings include Linlithgow in West Lothian, Stirling Castle and Holyrood Palace in Edinburgh and Falkland Palace in Fife. They have long and complex histories and their Renaissance influences were often just part of their historical development. Other architectural influences are also evident at each site.

The origin of Stirling Castle is as a defensive site on a craggy volcanic rock, but it was also a favoured place for royalty, so impressive palace buildings, a great hall and a chapel were all constructed there. Fenwick (1974: 128) suggests that "Stirling vies with Hampton Court as the first important building in Britain in which Renaissance details appear to any marked degree". The King's Master of Works was James Hamilton of Fynnart who brought French stonemasons over to carry out the work. The Palace block dates to c.1540 under James V and features sculptured heads and statues.

Linlithgow Palace in West Lothian was erected around 1500 by James IV, but with later alterations by Hamilton. It is a substantial site situated in a loch-side location and its architectural features include the royal arms above the entrance gate and numerous statues and sculptures. The building is now largely a ruin having been damaged by fire during the Jacobite rebellions in 1746.

17.1   Falkland Palace, Fife

Falkland Palace in Fife is on a much smaller scale and was occupied as a royal hunting lodge. Architectural features include the two turrets at the entrance and corbelled gables. Its Renaissance façade, dating to 1537–1542, has "large mullioned windows, wall shafts and portrait medallions, but it is mainly in the proportions and feeling that the Renaissance spirit resides" (West, 1985).

While these royal palaces were extended throughout the sixteenth century using Renaissance detailing, the domestic architecture of the Scottish nobility was also changing. In Part 2, the influence of a more settled and stable society on its architecture is discussed. It is in the transition in Scotland from defensive tower houses to more domestic-scale houses and then to mansions that there is evidence of greater prosperity and stability in society generally. These buildings were designed to be statements of prestige and status rather than to offer protection. Good examples include Castle Fraser (largely early 1600s) and Craigievar Castle (1626), both in Aberdeenshire, Glamis in Tayside.

Gradually, mansion houses were erected and began to display Classical influences. Traquhair House in the Scottish Borders began as a pele tower but through time the site was developed and it is now recognised as the oldest continually inhabited house in Scotland. The majority of the house as seen today was erected in 1642 by the First Earl of Traquhair. It displays elements of symmetry to the front elevation and as Fenwick (1974) describes, it changed from being a tall, vertical tower house to one much more horizontal in plan.

Chapter **18**

Baroque

High Renaissance, Mannerism, Baroque and Palladian might all be defined within the Renaissance genre of classically influenced styles. In Britain, Baroque was not adopted with the exuberance of Europe where Italy, France, Spain and Northern Europe took it up with enthusiasm. Eventually it morphed into the heavily decorated and articulated form known as Rococo. However, certain British architects did adopt it, including Wren, Hawksmoor and Vanbrugh.

Sir Christopher Wren (1632–1723) prepared plans for the rebuilding of London after the Great Fire of 1666 which destroyed much of the city. He was involved in the design of around 50 churches in London including St Paul's and became Surveyor General of the King's Works in 1668. Nicholas Hawksmoor (1661–1736) worked with Wren on St Paul's Cathedral and other designs. Sir John Vanbrugh (1664–1726) designed Castle Howard (1699–1726) which Curl (1999) says "was a virtuoso performance in the Baroque style, more Continental than English". He also designed Blenheim Palace in Oxfordshire for John Churchill, the First Duke of Marlborough following his victory against the French at Blindheim (Blenheim) after which the house is named. The building was started in 1707 and is monumental in its scale.

Baroque art and architecture were heavily influenced by the stylistic response of the Italian artist, architect and polymath Giovanni Lorenzo Bernini (1598–1680). Bernini amalgamated the art of the sculptor and the architect. The resulting Baroque style is identified by an exuberance of style and sculptural form. Baroque was the style that heralded the end of the period we identify as the Renaissance which was smoothly transiting into a general return to Classicism and classical architectural preferences during the eighteenth and nineteenth centuries.

Neo-Classicism might be considered a reaction to the excesses of High Baroque and Rococo where greater and greater added

18.1 St Paul's Cathedral, London

18.2   Baroque architecture in Venice

decoration and sculptural forms had become the norm. This transition to more subtle forms is tied in with a general revival of interest in the purer form of classical influences in art and literature, as well as architecture, particularly those of Classical Rome and Greece. It was the Age of Enlightenment, of the late seventeenth century and early eighteenth century that witnessed a sea change in thought, response and philosophy when classical values and ideals greatly influenced society and resulted in the great return to the purer forms of classical architecture evinced by the many great mid-eighteenth- and early nineteenth-century city redevelopments such as at Edinburgh, Bath, Newcastle, London and numerous other British cities and towns.

Chapter **19**

The Scottish Enlightenment

The adoption of the Italian Renaissance in Britain proved to be by slow transition. Early buildings demonstrating Italian influence simply adopted some of the elements but were fundamentally Gothic in style and origins. Gradually, by the seventeenth century a fuller understanding of the use of Renaissance styling emerged, with the eighteenth and nineteenth centuries being the greatest period when classically influenced styling was adopted. The impressively planned schemes of Edinburgh's New Town (James Craig, architect 1760–1820), Bath (John Nash), London, Newcastle and even smaller centres like Perth, owed their design philosophy to classical influences.

In Scotland, after the wars with the English between 1660 and 1670, there was a mood swing against the strict Calvinistic principles of the recent past. It was this mood swing that resulted in what is identified as the Scottish Enlightenment, although it should be emphasised that the philosophy of enlightenment was not confined to Scotland. Its influence was the slow transition from the Gothic and church-based society of the fourteenth through to the early sixteenth centuries with its sometimes dark and introspective thinking into a social philosophy influenced and engineered through the return to humanism germinated within the period we have identified as the Renaissance (rebirth). The term *enlightenment* is therefore, perhaps, apposite – the lifting of the curtains to let in the new light of humanistic thinking and philosophy – with the return to classically influenced values in art and architecture operating from the base influence of Classical Greece and Rome.

It was not only art and architecture that were influenced by the new thinking; banking and trade were also part of the process. It was a period of true and extensive change that heralded the move from the Middle Ages into the new era of the seventeenth century and beyond.

Fiscal thinking, in Scotland and elsewhere, was affected by the work of Adam Smith through his publication *The Wealth of Nations* (1776). He was probably the first person to recognise and promote the free market and consumerism. His thinking was to affect the financial world to the extent that, currently, his profile appears on English banknotes as recognition of that influence.

Many other enlightened Scottish thinkers encouraged the adoption of Classicism in art and architecture during the mid-seventeenth and early eighteenth centuries including Robert Adam, James Gibb and Colen Campbell. Philosophers like David Hume encouraged Britain to think differently and, of course, Robert Burns affected society through his poetry and, later, Walter Scott wrote stories and published books inspired by the new enlightened thinking but with a return to romanticism of a former heroic age. Engineering and science also had their parts to play and men like James Watt (1736–1819) made huge advances.

Scottish thinking, engendered by the Enlightenment, was spread around the world by the migratory patterns of the Scots, as Arthur Herman puts it in his book *The Scottish Enlightenment* (2003): "the Scottish diaspora, that extraordinary migration of ... Scots to every continent on the globe". This migration, in part enforced through the Highland Clearances, resulted in a significant impact, particularly on the New World.

However, Scotland, compared with England, was not a wealthy nation. A series of poor harvests combined with the disastrous attempts to set up a worldwide trading scheme through the Darien scheme resulted in virtual national bankruptcy at the end of the seventeenth century. With the Act of Union of the Parliaments with England in 1707 Scotland became more prosperous. It was this new money and confidence that facilitated the redevelopment of the New Town in Edinburgh together with large parts of Glasgow and Aberdeen, adopting the Classical architecture of the Enlightened Age. Similar processes were obviously affecting England, Wales and Ireland but, and it might be argued differently, the seeds of the new thinking had their germination in the Scottish Enlightenment.

The knowledge of influential people in society was also improved through travel as wealthy people were able to take the Grand Tour. They visited the Continent, travelling to Italy and Greece where they were exposed to Classical architecture as well as the great art of Italian cities. Even within Britain tours were undertaken. Dr Johnson famously toured Scotland, but others like Thomas Pennant and Daniel Defoe also left memoires and diaries of their time, giving fascinating contemporary accounts of their travels.

Chapter **20**

Georgian architecture

The eighteenth century was a period of dramatic change in society and architecture in Britain. The seeds had been sown during the seventeenth century, but it was during the Georgian era that we start to see changes in town planning and construction on a much more significant scale than before. The symmetrical influences of the Renaissance came to fruition in the eighteenth century in the Neo-Classical style. The period witnessed the building of some of Britain's greatest country seats as well as more utilitarian domestic architecture. There was also a greater emphasis on town planning for the first time and the wonderful crescents of Bath and Edinburgh exemplify this.

The most influential architects of the period were William Adam (1689–1738) and his sons John (1721–1792), Robert (1729–1792), James (1732–1794) and William (1738–1822). Robert was one of the family's most celebrated architects, having been inspired by the architecture of Italy where he spent four years during the 1750s. The interiors of Robert's buildings featured richly decorated plasterwork and fireplaces such as Lansdowne House, London (1765). Summerson (1996) describes Robert Adams' influence following his return from Italy:

> This was at a time when the fashion for building enormous Palladian country seats was just past its peak. Interest in bleak and massive exteriors was giving way to a desire for new standards of interior elegance and it was in the re-planning and decoration of existing houses that Adam made his name.

The Georgian development of Bath by John Wood (1704–1754) represents Classical architecture and town planning on an impressive scale. Wood's formal design included Queen Street (from 1728) although his son, John Wood the Younger (1728–1781), continued to develop Bath including the Crescent (1767). Youngson (1988: 70) states that the

20.1   Bath terraces and circuses

planning has similarities to the seventeenth- and eighteenth-century Squares found in London but what was original about Wood was

> his use of curving streets – circus and crescent – with all the houses joined together in what sometimes forms a single palace façade. This multiple house-building, the laying out of a whole street as a unit, was saved from dullness and wearisome uniformity by the elegant and varied detail in the architecture, coupled with the feeling of movement and variety given by the curve of the streets.

The use of curves was then emulated in other planned Georgian developments.

In 1707, following Scotland's financially disastrous collapse of the Darien colony, the Union of the Parliaments was enacted in the Act of Union. However, this did not bring peace between the two countries. Indeed, the considerable resentment at the Union encouraged the Jacobite uprisings of 1715 and 1745 which meant on-going battles with England until Bonnie Prince Charlie was finally exiled in 1746. Large tracts of land owned by Jacobite sympathisers such as the Earl of Perth were forfeited to the Crown and then managed by the Commissioners for the Forfeited Estates. However, they brought much-needed change and improvement to towns and villages, including new agricultural practices and construction of more permanent stone and slate houses. The arrival of a more settled period and increased prosperity also encouraged development in Scotland's cities and notably the building of Edinburgh New Town.

It was recognised that Edinburgh could no longer work within the restrictive confines of the existing city and that a radical new approach was required. The steep Royal Mile was heavily overcrowded with tenements several storeys high, poor sanitation and the Nor' Loch had become heavily polluted and smelled extremely unpleasant. In an attempt to rectify this, in 1752 a pamphlet, "Proposals for carrying works on certain Public Works in the City of Edinburgh" was published which stated that the aims were:

> To enlarge and improve this city, to adorn it with public buildings, which may be a national benefit, and thereby to remove, at least in some degree, the inconveniences to which it has hitherto been liable, is the sole object of these proposals.
>
> (Youngson, 1988)

A competition for the design of Edinburgh's New Town was won by James Craig and construction started on the North Bridge in 1765, leading from the existing town to where the New Town would be located. In July 1767 James Craig's plan was finally adopted. This involved a completely new system of streets and buildings based in a grid layout with fine houses and buildings in the Classical style, served by a new drainage system that involved draining the foul-smelling Nor' Loch into which the Old Town's drainage discharged. The original location of the Loch is now the site of the Waverley

20.2   Bronze of Edinburgh New Town layout

20.3   Great King Street, Edinburgh

Railway Station. The New Town was further extended during the period up to about 1820, resulting in the layout that we now see in Edinburgh and centred around the three main streets of Craig's original scheme of Princes Street, George Street and Queen Street, with two green open spaces or Squares at either end – Charlotte Square and St Andrew's Square.

Craig's original layout had George Street, not Princes Street, as its main thoroughfare. Because of its enthusiasm for Classical Greek architecture, Edinburgh was dubbed, the "Athens of the North". This is exemplified in the construction of the National Monument on Calton Hill, the exterior of which was to have been an exact copy of the Parthenon in Athens. Designed by William Playfair, work started in 1822 but by 1830 funds had run out and it still lies in its uncompleted, folly-like state, high above Edinburgh's streets.

Although Bath and Edinburgh were the inspiration for others to follow, schemes on a more modest, but no less significant, scale are evident in numerous towns and cities. In Perth, Robert Reid, architect of Edinburgh's Second New Town, designed part of the New Town in the 1820s. The medieval burgh which had been confined by its city walls broke out of its confines with the building of Smeaton's bridge over the River Tay in 1771. This allowed the town a new freedom to develop and it has some impressive palace-fronted buildings which look over the open lands or Inches, towards the River Tay. McWilliam

20.4   Rose Terrace, Perth

(1975: 127) states: "The new face of Perth looked outwards, enjoying the view, defining the town, and presenting to the approaching traveller a happy collusion of natural and man-made which is unique in Britain."

Chapter **21**

Georgian terraced and smaller houses

In addition to these large formal schemes, Classical ideals were applied to buildings of all types and sizes. The eighteenth century and early nineteenth century witnessed the development of domestic architecture in Britain, particularly with the greater availability of bricks for construction. Woodforde (1985: 12) states:

> Georgian domestic architecture was the first kind to rise above the regional variations of vernacular building, none of its house-fronts, at least, conforming to any pattern but that of the Italian Renaissance. A classical house of granite in Lancashire looks very much like a classical brick house in Kent, though the first may be 75 years younger than the second.

The Classical design of Georgian townhouses is plain brick walls with stone or brick dressings and with evenly spaced sash and case windows with small panes of glass often in a 12- or 18-pane arrangement. The proportions of the building are of crucial importance, with symmetrical and repetitive façades adopted to give refinement and simplicity. Gradually roofs became hidden behind parapets as they became less steeply pitched. The entrance door is typically panelled with perhaps Classical detailing surrounding it such as pilasters and a cornice. Gradually through the Georgian period the glazing bars became lighter, particularly with improving techniques in glass production. Although quite plain and perhaps even repetitive, Georgian townhouses can be very elegant. Some were undoubtedly derived through the use of pattern books which speculative builders used, and these allowed a dissemination of ideas and designs.

The introduction of building regulations had a significant impact on building design in the Georgian period. These were primarily aimed at reducing the fire risk in urban areas, controlling use of flammable materials like thatch or timber. Woodforde (1985: 46) states

21.1   Georgian brick terraces, Exeter

that in a London Act of 1708, windows were to be recessed rather than flush with the wall and states that this also affected the design of window shutters:

> The fire hazard caused protective window shutters to be fitted inside most houses, and not outside as on the Continent. These shutters for downstairs rooms, either neatly folding or moveable up or down on the sash window principle, needed to be strong as well as attractively joinered.

The Great London Building Act of 1774 laid down specific regulations in its code of structural requirements for foundations and walls. However, failures were still a problem because of the approaches to construction. For example, sometimes there was a lack of a bond between the outer and inner brick skins, and with the inner skin supporting joists it tended to sink, or with Flemish bond sometimes the brick headers were only half and not full bricks, resulting in a lack of bond between the two walls. More serious problems were caused by the use of "bond timbers" which were used in walls for strengthening. As the wood shrinks or rots there is uneven settlement, sometimes of considerable magnitude. While collapses were not unknown, numerous examples of these buildings still stand today so an understanding of how they were constructed is of critical importance.

Extensive rebuilding was not confined to the urban areas. Increased prosperity during the eighteenth century encouraged the rebuilding of many old rural cottages in Scotland. This was partly due to improved farming practices such as enclosing fields and improved approaches to agriculture, partly initiated by the Commissioners for the Forfeited Estates following the Jacobite uprisings. It was the period of the Agrarian Revolution (1760–1820). At the end of the eighteenth century a statistical account of Scotland was prepared by all the church ministers of Scotland's parishes. The detailed entries in this account, dating to the 1790s, note that in the preceding decade or two there had been considerable improvements in construction. An entry for Little Dunkeld in Perthshire describes the changes:

> Landlords are beginning to collect weavers and other handycraftsmen into small villages, where they are accommodated with neat dwelling houses, and each of them with an acre or two of land, to afford them the benefit of a milk cow and some other comforts of agriculture, without being too much hindered, by the labours of the field, from a vigorous application to their respective trades.

There is therefore a combination of improvements in social and economic conditions as well as in architecture. Improvements in construction are attributed to the availability of lime, slate and stone to build better and more permanent dwellings than had previously been the case in many small Scottish and English villages. These houses were generally symmetrical in design and the basic plan form is found throughout the country, although varying in size. Three windows at first-floor level and two windows and a central door at ground level is a common design found in cottages and farmhouses.

Some "model villages" were planned by local landowners who were seeking to improve their estates both aesthetically and financially. Better housing combined with improved agricultural practices offered improved income for the landowners. Sometimes there was an existing village and the villagers were moved to a new site such as at Fochabers in Moray where the village was moved to accommodate the new Gordon Castle in 1775 and Inverary which was founded in 1743 by the Duke of Argyll when he rebuilt his castle. Other planned villages or extensions to existing towns' regular layouts and house styles include Pultneytown, Wick (1787, Thomas Telford), Kenmore in Perthshire (1760, Earl of Breadalbane) and Tomintoul (Sir James Grant). McWilliam (1975: 88) states that there were over 200 planned villages built in Scotland during the eighteenth and early nineteenth centuries, and that they were planned in both the economic and the formal sense. He says that they "followed eighteenth century principles" of design and that "even after 1800 they tend to bear the solid stamp of eighteenth century good sense". This is evident in some of these villages today where a unified paint scheme, sometimes of black and white as is found in Kenmore and Inverary, adds to the character, townscape and unity of the village and is indicative of the original planned scheme.

The Commissioners for the Forfeited Estates also initiated many planned villages across Scotland during the eighteenth century. There were four planned villages on the forfeited Perth estate: Strelitz, Borelandbog Park, Benniebeg and Callendar. These were in part created for soldiers demobilised from Highland Regiments following the Jacobite uprisings. However, considerable difficulties were encountered in the setting up of these villages and the Commissioners eventually abandoned the idea of totally new villages. Benniebeg, for example, located between Muthill and Crieff, was abandoned and then flooded by Lady Perth to form the Pool of Drummond in 1785. Despite this failure, improvements continued at many other existing establishments, not least Crieff, where industry, craftsmen and the hotel trade were assisted by the Commissioners. In total 255 new settlements or modified existing sites were created and as Adams (1978: 61) states:

> The new village bore witness to comprehensive planning, with a spacious main street often opening out into a square in the centre of the village. When expansion took place a grid pattern was adopted, and the geometry was reinforced by the regular layout of house plots, gardens and lanes. Houses

21.2 Symmetrical house, Muthill, Perthshire, built 1812

21.3 Dunning, a late eighteenth-century village planned around formal squares and open spaces with houses straight onto the pavement and no front gardens

were built directly onto the street, partly to discourage the habit of keeping a dunghill at the door; also the absence of front gardens gave the village a stronger "urban" appearance than its size and function merited.

Adams goes on to point out that as much as a function of aesthetics, planned villages had an important economic basis. Landlords' rental incomes were enhanced by the higher density of housing and farm rents were increased to reflect the locality of a new market. Linen manufacture was also encouraged so that these new settlements had a sound economic base. In 1784, the Board for the Annexed Estates was discontinued (Smith, 1982). The Perth Estate was restored to the Drummond family under a special Act of Parliament in 1785 (Jamieson, 1993).

Chapter **22**

The Picturesque Movement

Inevitably there was a reaction against the strict proportions and regularity of Classical architecture. This led to the development of the Picturesque Movement at the end of the eighteenth century. The Picturesque Movement went against the formal ideals of Classical symmetry and introduced a freer approach with asymmetrical buildings set in gentle rolling countryside. Associated with the great houses of the eighteenth century were their gardens and landscapes with vistas, open aspects and Elysian narratives.

Although the landscapes in which these buildings were set appeared to be natural they were actually heavily contrived, although also highly successful. Vistas, water, trees and parkland were combined to create a perfect English landscape which the house could oversee. Lancelot "Capability" Brown (1716–1783) was a highly influential landscape architect who created apparently 'natural' landscapes for grand country seats. His works include Nuneham Park, Oxfordshire (1778–1782) and Croome Court, Worcestershire (1751–1752). Summerson (1986: 92) describes his effect on the English landscape, stating:

> Between 1750 and 1780 nearly all of the great English parks passed through the hands of Lancelot ("Capability") Brown who swept away the avenues and parterres and remodelled the grounds according to a vision of his own. This vision crystallised in a formula which embraced the new planting of clumps and belts of trees combined with the creation of new contours and artificial serpentine lakes.

Humphrey Repton (1752–1818) and John Nash (1752–1835) were both important architects of the Picturesque Movement following Capability Brown and worked together until 1802. Their approach was slightly different to Brown in that they sought to 'improve' the existing landscape by removing any obstructions and working with what was

22.1   Fountains Abbey, Yorkshire

already in existence rather than the major recreation of a landscape that had never existed in the first place (Summerson, 1986). Nash designed many villas, including the Italianate Cronkhill, Shropshire (*c.*1802) as well as Regent's Park in 1811. These landscape architects were influenced by the Classical scholars of architecture of the Renaissance and were in search of the Vitruvian ideal.

22.2   Ossian's Hall and Bridge, The Hermitage, Dunkeld, Perthshire

Sometimes the Romantic ideal was taken to extremes with the building of ruins and follies, sometimes hidden in order to surprise the visitor. The Hermitage in Dunkeld, Perthshire was originally designed as pleasure parkland for the Duke of Atholl's Dunkeld House. This large woodland area is situated adjacent to the River Tay and the Falls of Braan. This landscape was enhanced with the planting of exotic tree species and the construction of unusual buildings. The buildings included a folly, renamed Ossian's Hall in 1782. It originally had a mirrored interior for theatrical effect, a stone bridge crossing the falls and Ossian's Cave, a man-made "cave". These were erected in the 1780s as part of the enhancement of the visitor experience (Haynes, 2000).

Chapter **23**

Regency architecture 1790–1840

Although the Regency period is generally regarded as a continuation of the Georgian period of the eighteenth century, the circumstances in the early nineteenth century were different and this had a direct influence on the architectural style of the period. The Napoleonic Wars had caused shortages which resulted in a doubling of building costs in the two decades between 1790 and 1810. Increases in prices for land and building materials, together with increased pressure for housing in urban areas, resulted in lowering of building standards (Woodforde, 1985). For example, floors sagged due to insufficient timbers, although this tended to be hidden by the use of lath and plaster ceiling.

Externally, cheaper construction was hidden by the use of stucco. Advances in brickwork production meant that bricks were cheaper and more readily available compared to stone, which had become increasingly expensive. However, the appearance of stone could be achieved through the adoption of stucco which could be cut into with false joints to resemble ashlar work or string courses. External stucco was typically made with very fine sand and white carrara marble to achieve a glistening white finish. Stucco could also be sanded and polished to achieve a very fine finish. It was particularly fashionable in the early nineteenth century and was notably used in seaside resorts like Brighton, Llandudno and Hove. This lent these buildings respectability and made them fashionable, unlike the Georgian brickwork which was considered to be inferior in appearance. This was therefore a deliberate act of deception in presenting buildings as stone where they were in fact constructed of brick. This fashion encouraged the invention of artificial stone products like Coade stone which became popular during the nineteenth century.

The principles which had applied during the Picturesque Movement continued to exert an influence during the Regency period with the layout of parks, houses and streets like Regent Street in London. The

23.1   Stucco buildings in Exeter

garden and landscape was of special importance to the architects of the period as Pilcher (1947: 43) indicates:

> To get as close as they could to nature was, in one sense or another, the consistent ambition of the Regency architects. They had started by carrying the house into the landscape. They finished by bringing the garden into the house.

This involved new structures like conservatories which linked garden and house together. They also used rustic features like tree trunks to support porches. The Regency period also popularised the use of ironwork. Balconies featured on the front elevations of buildings, sometimes in ornate designs, although many were of standard designs.

The planned schemes of Bath and Edinburgh which had their origins in the eighteenth century continued to develop, and comprehensive town planning became a distinct aspect of Regency architecture. In Newcastle, builder Richard Grainger built Eldon Square (1824–1826) and the Triangle (1834–1839) and Grey Street, which according to Watkins (1982) has its inspiration in the Picturesque design associated with Regent Street. Here a shopping and office arcade were erected, a feature of the new move into commercial premises which also typified the period, partly attributed to the growth in banking and life assurance. Other commercial improvements included markets like Covent Garden in London and Argyle Arcade in Glasgow (1827).

Chapter **24**

Victorian period (1834–1900)

Queen Victoria ascended to the throne in 1837 and remained there until her death in January 1901. In terms of social, economic and architectural development, the Victorian era is one of the most dramatic periods in recent history for the United Kingdom. There were also wider geographical repercussions given the extent of the British Empire across the world and the consequent effect on the social and built structures of those countries that formed part of the Empire. Within the British Isles, there was a transformation from an essentially agrarian society to an industrialised and urban society in less than a century. The population rose dramatically, more than tripling in the course of a century. In 1801 the population was ten million, but had risen to almost 21 million in 1851 and then 37.5 million by 1901. This had profound implications for the built environment and the need to house people, provide places of work and centres for government and learning. The rapid rise in population and increase in industrialisation and ultimately urbanisation created opportunity for some, along with misery and squalor for others.

Architectural styles

The Victorian period extends over two-thirds of a century and witnessed a variety of influences over that time. The Classical architecture of the Georgian period remained popular but other styles like Gothic were revived and there became a great fashion for elaborate ornamentation and combining of styles. Notably, the architectural profession developed significantly during the nineteenth century and was no longer exclusive to a few key people. In 1834 the Royal Institute of Architects was granted a Royal Charter, helping to promote the profession. Local architects began to have an influence in their particular town or city and often made a significant contribution. Sole partnerships or family firms began to merge to create larger practices that could deal with the significant and increasing volumes of work available.

Despite the local influence of these individuals, there were also clearly those who were seen as making a contribution nationally. These include Augustus Pugin (1812–1852) and Alexander "Greek" Thomson (1817–1875), as well as writers such as the influential Ruskin (1819–1900) and William Morris (1834–1896).

While the Georgian period had focused on classical architecture, during the Victorian era the design influences were less confined. Styles from previous periods were revived, altered and combined which often resulted in elaborately decorated buildings where no particular style was evident. Architects battled with each other as to which style should prevail, Gothic or Classical, in the so-called "Battle of the Styles" and in the end many pursued their own chosen style. The Victorian architectural period is described by Summerson (1996: 108) as follows:

> The nineteenth century was very much concerned – over-concerned we may think – with historical styles. Classical buildings were continually being built but they always looked back, not merely to Greece and Rome but to nearly every succeeding phase of classical development, using the past as one glorious quarry of ideas.

Aston and Bond (1976: 191) state:

> Stylistically the Victorian period is confused. The main theme is revivalism, but the wider experience of foreign travel had introduced a bewildering variety of designs on which to draw. Gothic, Classical, Romanesque, Moorish, Byzantine, Egyptian and Chinese styles were all absorbed into the Victorian repertoire. Most architects and their patrons had little or no formal architectural training, and the results are sometimes unfortunate, with the simplest building becoming overloaded with a cumbersome mass of misunderstood and ill-used detail.

Greek Revival or Neo-Greek

This had its origins in the eighteenth century when publications like Stuart and Revett's *Antiquities of Athens* (1762) and Winckelmann's (1717–1768) various publications on Ancient Greek architecture were published. Architects such as William Henry Playfair (1790–1857) designed a number of buildings, particularly in Edinburgh, including

24.1   The National Gallery of Scotland, Princes Street, Edinburgh

the Royal College of Surgeons (1830–1832) and the National Gallery of Scotland (1850) with its Greek figures and temple-like appearance. Also in Edinburgh, Thomas Hamilton (1784–1858) designed the Royal High School in 1825 and the Royal College of Physicians in Queen Street in 1854.

The architect who is most famously associated with this style is Alexander "Greek" Thomson (1817–1875). He worked primarily in Glasgow where he designed numerous tenements, churches and commercial buildings of distinctive design. These include the Buck's Head building in Argyle Street, Caledonian Road Church, Egyptian Halls and Grecian Chambers. McFadzean (1979) states that Egyptian Halls, erected in 1871 for Glasgow businessman James Robertson, is the "most exotic" of Thomson's commercial designs and contains a series of ground-floor shops.

Gothic Revival

The Gothic Revival was led by Augustus Pugin (1812–1852). He was notably employed by Charles Barry to work on the Houses of Parliament and also designed Scarisbrick Hall (1837). In 1841 he published *The True Principles of Pointed or Christian Architecture* on his theories on Gothic architecture as the true style. He was a Catholic convert and heavily influenced by his religion and believed that Gothic was the true and honest style and that other styles, such as Picturesque, were dishonest.

Building types

The architecture of the Victorian period is therefore complex. Although some of the buildings erected during this time are at best exuberant and at worst unattractive, they make a significant contribution to the townscape of Britain. One of the most fascinating aspects of the period is the numerous new types of buildings constructed. All aspects of society during the nineteenth century changed, everything from the way people bought goods to the way they moved around the countryside. This necessarily involved the creation of completely new building types. These are now briefly outlined.

Public health

The Victorians introduced legislation to improve public health. Associated with that were new types of buildings such as hospitals and morgues. New cemeteries were created to cater for the dead, such as the Glasgow Necropolis. These large sites were designed to be picturesque places people could visit and wander around, similar to a park or garden.

The Victorians also introduced extensive drainage, sewerage and water supplies into towns and cities. The great rebuilding/construction of London sewers and river barriers by engineer Joseph Bazelgette (1819–1891) are a great example of the improvements in public services during the Victorian period. The rapid urbanisation meant that without such improvements the towns and cities would have become unbearable in terms of public health.

Education

The Education Act of 1870 encouraged the construction of new schools and institutes to improve the education of people.

Entertainment

More money and leisure time meant that the Victorians had an increased desire for leisure pursuits. These included theatres and music halls as well as public houses and hotels. Glasshouses like Kibble Palace in Glasgow were also constructed to use as places of entertainment. Seaside resorts like Blackpool flourished with the improved train and tram network. Blackpool Tower, which emulated the 1889 Eiffel Tower in Paris, was opened in 1894, complete with

24.2   Buxton pavilion and gardens

entertainments which included an aquarium, zoo, circus and Rococo ballroom. There was a move away from urban centres as havens of leisure (eighteenth-century polite society and the assembly houses of the larger cities and towns) to the spa and seaside resort.

Hydropathic establishments which were considered to be healthy and away from the dirt and unhealthy air of the cities were developed as tourist resorts. Towns readily accessible by the new train network centres included Harrogate in Yorkshire, Buxton in Derbyshire and Crieff in Scotland. These helped to develop new types of structures, large hotels and associated buildings to cater for tourists.

Commerce

There were changes in business opportunities with foreign trade and importation of materials like cotton, tobacco and sugar. Mills and factories were needed to process these materials, along with harbours and warehouses to deal with their transport and storage. In addition, shops were needed to sell goods and even shopping centres were created. Formal market halls where the selling was under cover and markets divided by type, such as meat markets or cheese and butter markets, gradually replaced the old street markets. These were created to prevent the nuisances caused by street markets which encroached onto the thoroughfare and were more difficult to regulate.

24.3   Forth Rail Bridge, Edinburgh

Transportation

As well as public health improvements, one of the greatest achievements of the nineteenth century was the huge improvement in the transport network. This included canals, viaducts, railways and bridges. Railway stations were constructed throughout the country and even located in the smallest rural settlement, allowing extensive transport throughout the country. St Pancras Station in London (built in 1868) is an example of the design of railway stations using glass and iron.

At times these structures pushed the limits of technological knowledge. In Scotland the Tay Rail Bridge was constructed in 1878 using wrought and cast iron, but it subsequently collapsed in 1879 as a train was passing over it. As a result the Forth Rail Bridge was constructed in steel and opened in 1890. It is a remarkable feat of engineering and is still in daily use today, although it has proved to be onerous in terms of maintenance.

Social reform

The Victorians were enthusiastic social reformers. They built prisons to house those who had committed crimes. Lunatic asylums were also constructed throughout the country to accommodate those considered to be insane.

Churches

Church building was also central to community life in the nineteenth century. With public worship attendance high compared to the twenty-first century, many churches were of significant size in order to cater for the large numbers of church members. Styles varied throughout the period, but the Gothic Revival led by Pugin was adopted as a suitable style for many churches throughout the United Kingdom.

Public buildings

Municipal buildings like town halls were erected to accommodate local government needs due to the increasing role which county and local councils played in running the affairs of the country.

24.4   St Giles, Cheadle, Staffordshire, designed by Augustus Pugin and known locally as "Pugin's gem"

Victorian housing

There were particularly strong contrasts in the different types of housing that existed. Large stone villas set in generous garden grounds were built for the rising middle classes, many of them having discovered new-found wealth through business opportunities – some of which were overseas in the British colonies. These new businesses included trades in tobacco, sugar and cotton, largely founded on the African slave trade. The houses reflected this opulence and wealth, being of high quality and permanent construction, and using dressed stone and slate, these were set in the new suburbs of towns and cities away from the soot and dirt of the industrial factories.

In strong contrast, the monotonously identical terraced housing of northern England situated in the lee of mills and factories housed the workers. The brick and slate "two-up, two-down" were rolled out across the country with minimal space and only a small backyard with no garden area. Stone was used for construction where locally quarried, such as in Lancaster, but brick, which was readily available and easily transported by the new networks of canals and railways, was adopted throughout. Industrial towns like Accrington, Blackburn and the Potteries of Stoke-on-Trent are of brick construction.

24.5   A stone villa set in generous grounds, Glasgow

24.6   Workers' housing in Stoke-on-Trent

24.7   Typical rear yards of "tunnel back" terraces, Fenton, Stoke-on-Trent

24.8   New Lanark workers' housing

High-density housing was used across the country for working-class people, many of whom had moved from rural areas in search of work. Some had been forced to move due to their particular local circumstances such as the Highland Clearances and the Irish potato famine. Regional differences emerged such as "back-to-back" housing in Leeds and tenements in Scotland. Notorious areas like the Gorbals in Glasgow epitomised the misery through lack of facilities and families crammed into one small room with a lack of ventilation or sanitation. As a result, disease was rife.

However, not everyone took advantage of the working classes and certain individuals sought to enhance their life and offer good housing and education. Sir Titus Salt (1803–1876) built the village of Saltire following his factory construction of 1853, while the Cadbury family built Bourneville and Robert Owen built New Lanark in Scotland. These were all leading examples of attempts to improve the lives of mill and factory workers.

Materials

Without the development of new materials during the nineteenth century, many of these buildings could never have been constructed. The development of cast iron, plate glass and other mass-produced materials facilitated the construction of large buildings. This is epitomised in the creation of the Crystal Palace for the Great Exhibition of 1851. This incredible structure was made possible by the combination

24.9   Gardiners Warehouse, Jamaica Street, Glasgow

of glass and cast iron, using over a quarter of a million panes of glass and standing at 2,000 feet long by 408 feet wide.

The subsequent lifting of the glass tax in 1845, which had previously penalised glass on a weight basis, meant that glass was now both readily available and cheaper. Although there was still some prejudice against the use of cast iron and plate glass, the Great Exhibition helped to make them fashionable and acceptable. They subsequently became used in all types of structures including railway stations, shop fronts, bandstands, shopping arcades and fountains. Great warehouses like Gardiners Warehouse in Jamaica Street, Glasgow of 1856 provided a wall of glass and cast iron. Numerous iron foundries became established in Scotland, the most famous being Walter Macfarlane's Saracen Foundry in Glasgow which became one of the biggest exporters of cast iron across the world. Customers could choose items from the foundry catalogues and piece together the building of their choice.

The industrial scale of production of building materials like slate, brick and stone together with the ready availability of railway and canal transport offered tremendous opportunities for construction. However, the downside of this availability was a steady erosion of local character. The transportation of Welsh slate throughout the country meant that local materials were not used. Architects were increasingly adopting ideas from larger centres and, as a result, local vernacular traditions began to decline. This was following a steady trend in cities where traditionally used materials like timber and thatch were being banned through local byelaws in an attempt to reduce the risk of fire.

Aston and Bond (1976: 191) state:

> The quicker, cheaper bulk transport offered by the railways finally disrupted the boundaries of local building materials. Welsh slate, knapped flint, all types of stone, terracotta and ornamental polished granites and marbles could be carried to any part of Britain and used in any imaginable combination. The removal of the brick tax in 1850 and the mechanisation of the industry made it possible and worthwhile to produce enormous quantities of bricks which were almost identical in size, colour and texture.

The nineteenth century saw considerable expansion in all fields – domestic, commercial, public, religious and industrial. All sectors saw new types of construction as engineering and new products were introduced. However, towards the end of the period there was something of a backlash and a rejection of the mass production of materials. From this, William Morris and the Society for the Protection of Ancient Buildings led the Arts and Crafts Movement that advocated a return to hand-made and "honest" materials. This then took the Victorian era into the early twentieth century and many of these ideas were to influence the architects of the Edwardian period.

Conservation attitudes to Victorian buildings

Although we now consider many Victorian buildings to be of importance, this has not always been the case. Victorian buildings are a dominant building type for many British towns, representing a period in history when there was great expansion and change as society became urban rather than rural-based. Many of our churches, public buildings, hospitals, prisons and educational establishments have their origins in the nineteenth century, though they may have subsequently been altered. Their sheer numbers and dominance perhaps makes them seem less special than older or more individual styles of buildings. Indeed the 'run of the mill' nature of them means we do not always see them for what they are and do not recognise them as worthy of conservation.

However, closer inspection identifies them as an integral part of our historic townscape and one where materials like ceramic tiles, stained

24.10   Mill buildings in Leek, Staffordshire, England

glass and cast iron combine with sandstone and slate to provide impressive homes and public spaces.

Some of the buildings that have been least regarded in the past but are now seen in a new light are former industrial mills. The fortunes of these mills, the cradle of economic prosperity and work for many regions, changed in the twentieth century as the importance of the British Empire declined and cheaper cloth became available as imports from abroad. The closure of these mills meant that they no longer had a viable use and so they declined rapidly through lack of repairs, vandalism and robbing of valuable building materials. However, slowly it has been recognised that these mills play an important part in the social, cultural and architectural history of this country and are being revitalised.

New Lanark, located southeast of Glasgow, is a good example of this. A substantial site, it was founded on the Falls of Clyde by David Dale in 1786 and managed 1800–1825 by his son-in-law, the philanthropist and social thinker, Robert Owen (1771–1858). He built a school, housing and a church for his workers and was one of the first mill owners to have an interest in the social welfare of his employees.

During the twentieth century the mill declined, closing in 1968 and falling into serious disrepair. It was subsequently renovated by a Trust

24.11   New Lanark

which was established in 1974 and work is still on-going to renovate all the buildings. In 2001 it was declared a World Heritage Site demonstrating the considerable and international significance of this site. It is now an interesting mixture of tenanted houses, a museum with associated shop/café and a new hotel and conference centre. There are plans to build a swimming pool and associated facilities to meet the increasing demands of visitors. The desire to meet these needs causes conflict in trying to sensitively manage a World Heritage Site in an appropriate and sustainable way. Compromises are sometimes necessary in order to ensure an economically viable future for such a site.

Chapter **25**

Arts and Crafts Movement

The Arts and Crafts Movement was a revolution in architectural philosophy which arose in England in the latter part of the nineteenth century, led by William Morris. It placed emphasis on the materials themselves and the individual craftsmen rather than mass production of inferior goods. It also had a social and political dimension as Morris was a social reformer as well as an architect.

The movement has its roots earlier in the nineteenth century in the work of Augustus Pugin who was instrumental in the Gothic Revival of the early Victorian period. His buildings included Scarisbrick Hall, Lancashire and St Giles, Cheadle, as well as having been involved with Charles Barry in the winning design for the Houses of Parliament. He advocated constructional honesty, believing that architecture could be morally good or bad (Watkins, 1979). As a Catholic convert, he viewed his subject very much from a Christian standpoint and he felt strongly that buildings should be honest, which meant using locally available materials. This principle was to become a central doctrine of the Arts and Crafts Movement. The influential writer and evangelist John Ruskin also believed in constructional honesty and the use of handcrafted materials.

From these beginnings in the theories of Pugin and Ruskin, the ideas were carried forward by a group of painters, poets and critics who were known as the Pre-Raphaelite Brotherhood. Established in 1848, their outlook, like that of Pugin and Ruskin, was essentially Christian. Adams (1987) states:

> The Arts and Crafts Movement in Britain emerged, in part, from the works of the Pre-Raphaelite Brotherhood, a group of dissident artists who rejected the conventional artistic opinions of the academic establishment and sought inspiration in the arts, and later in the crafts, of the Middle Ages.

The central figures were Dante Rossetti and with his brother William, John Millais and William Hunt. Their periodical *The Germ*, came to the attention of William Morris and he, along with Edward Burne-Jones, joined the Brotherhood. They set up an artist's studio but Morris' interest in painting was short-lived and in 1856 he started work in the architect's office of G. E. Street where he met Philip Webb (1831–1915). Both men were heavily influenced by Ruskin's writings and developed his ideas in their years working together.

From 1859 to 1860, Webb built The Red House for Morris which is one of the first examples of Arts and Crafts architecture. Webb attempted to create a styleless vernacular (Watkins, 1979) with its simple design of plain red brick walls with a tiled roof, plain whitewashed interior and exposed timbers. Houses of the Arts and Crafts period display attention to detail and may feel vernacular in inspiration, making use of materials like timber and tiles. The small details like joinery finishing and ironmongery make these buildings unique, even if they may appear outwardly somewhat understated.

As well as architecture, the Arts and Crafts Movement involved the development of interiors including furniture and wall coverings. The establishment of the firm Morris, Marshall, Faulkener & Co. in 1861 was a crucial step in the development of these decorative arts that were to become so central to the movement. The firm aimed to promote hand-made products such as furniture, stained glass, wall coverings and textiles which were in contrast to the mass-produced

25.1   Skirling House, Biggar, Scotland

goods of the Victorian factories. They received many important commissions including the Dining Room at South Kensington Museum and the Tapestry and Armoury Room at St James's Palace. The firm then became synonymous with decorative artefacts of fine workmanship and natural beauty.

Morris' ultimate aim was to improve everyday products by making them beautiful and finely crafted. He saw factories as soulless places which produced low-standard products which were demeaning to the workforce. He sought to restore dignity to workers through his form of creativity. However, the problem with this logic was that the goods he was producing were expensive because they were handcrafted and of very high quality. As a result, the very people he wanted to help were unable to afford them and indeed, his products were mainly commissioned by the type of people who owned the very factories he detested (Adams, 1987). This paradox of the Arts and Crafts Movement was never satisfactorily resolved by Morris and in an attempt to overcome this he moved into politics in later years.

It is evident that Morris was an extremely influential figure who was driven by his socialist ideals to try to change the face of England through the Arts and Crafts Movement. His firm also inspired many architects and artists to carry on through the formation of a number of craft guilds such as the Art Workers Guild which was formed in 1884. The establishment of the Arts and Crafts Exhibition Society, an organisation which set up an exhibition every three years, in 1888 further enhanced the profile of the Arts and Crafts Movement.

The guiding principles of the Arts and Crafts Movement were rooted in the theories put forward by Ruskin and Pugin earlier in the nineteenth century. Their ideas were Christian-based and they sought an environment where architecture was honest, using local materials and handcrafted by individuals, not machines. The architects who followed the movement sought inspiration from a nostalgic and romantic world based around the architecture of the Middle Ages. They viewed this period in history as a time of simplicity and high moral values. It was felt that the workers were skilled craftsmen who took a pride in their work and that profit was not the sole motive. This utopian ideal was in strong contrast to the factories and overcrowded slums of the late Victorian period. Architects such as Morris sought to improve people's lives through the decorative arts. His vision of restoring dignity to the workforce, however, was an impractical one and indeed his very attempts at this in reality created a paradox which he could not resolve.

The Arts and Crafts Movement represented a period of change in our architectural heritage when many of the architects of the time were trying to reject the forces of industrialisation in favour of handicrafts and individualism. They attempted a return to the Middle Ages, a period they fondly viewed as honest and dignified, by revitalising the decorative arts and crafts of that period. Their architecture was concerned with the vernacular and promoted the use of local materials. It also attempted to create individuality rather than a specific style and in doing so various historical features came to be used within a single building. The whole movement was inextricably linked with social purpose and it was the aim of the architects involved in this revitalisation of medieval arts and crafts that it would serve to improve society as a whole. However, although the principle of using the Arts and Crafts Movement as a vehicle for social change was undoubtedly unsuccessful, the art and architecture which was developed during this period produced a wealth of innovative ideas that still inspire us today.

Chapter **26**

The late nineteenth and early twentieth centuries

The Edwardian era saw a general continuation of the styles that had prevailed at the end of the nineteenth century. Service (1975: 485) describes the period as "dualistic", combining a nostalgic golden image of wealthy country house owners and garden parties with a period of "intense social strife" recognised in the Suffragette Movement, the rise of the Trades Unions and the Miners' Strike of 1911.

Architecturally, the period was not straightforward. There were many influences, including a revival of English Baroque which was reminiscent of leading architects like Wren and Vanburgh. The period did, however, see the emergence of new design ideas, particularly in interiors with the Art Nouveau and Art Deco Movements.

Art Nouveau *c.*1888–*c.*1914

The Art Nouveau movement was a style of architecture and related decorative arts at the end of the nineteenth century. It was found in Britain, in parts of Europe and also in America, but varied geographically. It is defined by Curl (1999) as a European style which featured "asymmetrical composition, attenuated blooms, foliage roots and stems with sinuous flowing lines, as though floating in water; the dream maiden; stylised rose bowls; intertwining plant forms; and indeterminate whiplash curved tendrils".

It particularly utilised stylised flower motifs. Victor Horta (1861–1947) from Belgium rejected the historical styles which had been widespread in the nineteenth century for ornamental motifs in metal and glass which were inspired by plant forms (Roth, 1993). The style was named after the shop "Maison de l'Art Nouveau", owned by art dealer Siegfried Bing (1838–1905), which sold items of a new design, some of which were oriental in inspiration. It was also known in France as the "Style Moderne" (Curl, 1999).

In Scotland, architect Charles Rennie Mackintosh (1868–1928) introduced radically new styles in interiors and architecture. He trained with Honeyman and Keppie, architects from 1889, becoming a partner in 1904 to make the firm Honeyman, Keppie and Mackintosh. Mackintosh took his inspiration from the past – from Scotland's vernacular tradition, but he also embraced modern technology including steel, concrete and other man-made materials (Grigg, 1987). His designs were therefore totally new and inspiring, not least because he also designed interiors including furniture and light fittings.

He was commissioned to design a number of houses including Windyhill, Kilmacolm (1900) and Hill House, Helensburgh (1903), as well as commercial premises including Scotland Street School (1906) and Miss Cranston's Tearooms in Glasgow. One of his most famous designs, however, is for Glasgow School of Art, built in two phases: the east wing 1897–1899 and the west wing 1907–1909. Howarth (1977) states that "[i]t was and remains Mackintosh's most representative work, and it is undoubtedly his most important contribution to the New Movement". McWilliam (1975: 175) describes it as a "pioneer building" which is a "lonely peak of intelligence and refinement".

26.1   Hill House, Helensburgh

Mackintosh was a pioneer of the Modern Movement at the turn of the century and inspired the Art Nouveau style. Grigg (1987) states:

> Mackintosh found his inspiration in the vernacular buildings of the countryside. It is in the old cottages and details of village churches, and especially in the Scottish castles and tower houses, that fill his sketchbooks and are the vocabulary of his architecture. All his buildings are deeply rooted in the past, they have the quality of belonging to rather than being arbitrarily imposed upon the landscape that characterises vernacular architecture, yet Mackintosh's creativity is never hindered by nostalgia.

His wife Margaret MacDonald, a jewellery designer, worked closely with him and she was particularly influential on the artefacts and interiors. They both collaborated with Margaret's sister Frances and Herbert MacNair and they were highly original designers of buildings and interiors.

Mackintosh, though undeniably extremely talented, gained little respect in his own country and his work went largely unappreciated until after his death. His buildings were innovative and their interiors striking and unusual. Buildings like Hill House in Helensburgh typify this. Designed for publisher Walter Blackie, it has a rather severe appearance of grey harl but with vernacular inspiration of a

26.2   The Mackintosh Building, Comrie, Perthshire

pepper-pot tower, multi-paned windows and slated roofs of varying angles and depths. Although he had a good relationship with Blackie, his lack of compromise where his designs were concerned made him unpopular with many of his clients. However, he did design a shop and house in Comrie, Perthshire for friends, the Macphersons following a fire at their building. It is a very understated building but has a fully fitted out draper's interior.

In 1913 he left Scotland and moved to England but he achieved little success other than 78 Derngate, Northampton (1919) and in 1923 he moved to the south of France where his talent was more readily appreciated. Despite the high praise which he received on the Continent he died in 1928 in London in total obscurity, and it was only much later in the twentieth century that his unique and inspirational designs became recognised for their outstanding significance.

Edwardian architectural styles

Known as the Belle Époque (beautiful era), the Edwardian era relates directly to the period of the reign of King Edward VII (1901–1910) but is affected by events both at the end of the nineteenth century and into the period of the First World War (1914–1918). The war fundamentally and permanently changed social attitudes in Britain. Previously the class structure was rigidly adhered to; lower classes and particularly women were politically disadvantaged. With the end of the war Britain became a very changed place socially, with an increase in interest in socialism and women's suffrage. It was a period of great middle-class development.

The general 'feel' of the arts and architecture was one of change from the industrial influences of the Victorian period and its revival of Gothic styles into a much more introspective period of quiet revolution via the Arts and Crafts and Art Nouveau movements.

The Edwardian period was particularly focused on monumental architecture, carried forward by the corporation architecture of the Victorian period and its visually dominating town halls and public buildings. There was a return to Classical styles, the term Edwardian Baroque or Neo-Baroque being redolent of this influence.

The period demonstrates a complete mix of stylistic response, a truly eclectic pattern in architecture from the monumental of Edwin Landseer Lutyens (1869–1944) to the more domestic scale of Charles Voysey (1857–1941). Within Edwardian architecture, and demonstrative of the eclectic nature of design during the period, we also have the Art Nouveau style of Charles Rennie Mackintosh (1868–1928) and the Arts and Crafts influences of William Morris (1834–1896).

The early twentieth century was a period of change, both socially and in art and architecture. Education for the masses was a social priority. The 1902 Education Act took power away from the school boards established under the 1870 Education Act, and placed education and its control at a local level under Local Education Authorities (LEAs). This stimulated a significant development in school building and many new secondary schools and colleges were built or expanded at this time as well as numerous public lending libraries. In Scotland many of these libraries were funded by the philanthropist Andrew Carnegie (1835–1919). He moved to America but encouraged the pursuit of education and learning, which he firmly believed in the importance of, in Scotland through the funding of public lending libraries. The library in Montrose is in an Edwardian freestyle and was erected in 1905, designed by Edinburgh architect J. Lindsay Grant. Typical of many of these libraries, the interior is finished to a high standard with tiled floors, stained glass and high-quality joinery fittings.

A new approach to housing was inspired by the Garden City Movement which had Ebenezer Howard as its leading exponent. In 1898 he set down his philosophy in *Tomorrow: A Peaceful Path to Real Reform*. In 1899 he established the Garden City Association (Hockman, 2002). New Edwardian suburbs such as at Bedford Park in London were examples of the new style in domestic architecture influenced by the Arts and Craft Movement of the late nineteenth century. Bedford Park was commenced in 1875 and was completed in 1914. In its later

26.3   Public lending library, Montrose (1905)

26.4   Houses in Bedford Park, London

period of development it is very typical of the styles in domestic architecture of the Edwardian period. Bedford Park is also very typical of the garden cities and suburb developments such as Letchworth and Hampstead Garden. These 'garden' developments were directly influenced by the Aesthetic Movement of the late nineteenth century, seeking to return to a rural rather than urban idyll. It was a movement of rejection of the rampant urbanisation of the Industrial Revolution during the Victorian period.

The Edwardian period also had its share of the monumental in architecture and examples may be demonstrated by Edwin Lutyens, Richard Norman Shaw (1831–1912) and Giles Gilbert Scott amongst others. Their designs for post-First World War memorials and many large-scale hotel, shopping, banking and local authority buildings exist in many of our towns and cities nationwide.

Edwin Lutyens was a leading designer for medium-sized as well as large-scale country houses and is particularly well known for his liaison with the garden designer Gertrude Jekyll in creating houses and gardens such as The Deanery Gardens at Sonning in Berkshire and Munstead Wood, Surrey (1896). Championed by *Country Life* magazine, his designs were sought after by the nouveau riche ("new rich"), whose wealth had been expanded by the Industrial Revolution of the previous century. Lutyens also designed new second or holiday retreats for the wealthy Edwardian families including Overstrand Hall in Norfolk, commissioned by Lord and Lady Hillingdon, and The

26.5   Overstrand Hall, Norfolk

Pleasuance, commissioned by Lord and Lady Battersea. The grounds here were designed by Gertrude Jekyll. However, Lutyens is probably most famous for his designs for and planning of New Delhi, India, constructed during the period 1912–1931 when it was constructed to replace the former capital, Calcutta.

This epoch also witnessed a huge expansion of building, reflecting the still relatively cheap cost of building procurement, all too soon to be shattered by the cataclysmic impact of the First World War. It was the great period of the department store, grand hotel, mansion flat, large country houses, hospitals, schools and universities, as well as buildings to service the new power source, electricity. Office buildings took on a new dimension, housing large numbers of people to service an ever-increasing and expanding administrative focus. Buildings such as the new Norwich Union Offices in Norwich, England by G. J. Skipper are now regarded as the finest examples of Edwardian corporate architecture.

Improvements such as the invention of the tungsten filament bulb in 1910 meant that electricity became particularly influential in the design of buildings as availability of natural light was no longer such a governing factor. This was particularly useful for large buildings such as department stores, hotels and offices. It saw the expansion of the motor car, the telephone and the London Underground and was probably the last great period of expansion in Britain after the Second World War.

26.6   Norwich Union Offices, Norwich

The Edwardians experimented with the use of reinforced concrete for buildings, a material that was already being widely used in America. Buildings like Selfridges department store, which was built to an American model in 1907–1909, changed attitudes to design and the use of modern materials. However, Fellows (1995: 59) states that it tended to be hidden by more conventional materials like terracotta or stone, although its use was not necessarily totally disguised:

> The trappings of the Edwardian Baroque were often retained, but the fact that reinforced concrete had been used could be seen externally either in the bulk and diversity of the building or in the regularity of the framing coupled with the size of the window openings.

An example of this is Cardiff City Hall by Lanchester, Stewart and Rickards. It has a steel and concrete frame faced with ashlar stone.

Fellows (1995: 83) summarises the period, stating:

> Certainly, the Edwardian architect was extremely fortunate professionally, and the need for many new buildings in a variety of different types must have provided stimulating challenges. At the same time, the availability of new technology allowed for greater flexibility in the construction and servicing of buildings. A big scale was possible, where appropriate, as well as the opportunity to open up internal spaces, creating Baroque swagger, both externally and internally.

Art Deco and the inter-war period

Art Deco emerged around 1908–1912 although it is commonly associated with the period after the First World War. It used fine materials and lavish ornamentation and was an extremely complex style. Duncan (1988: 7) describes it as "the last truly sumptuous style, a legitimate and highly fertile chapter in the history of the applied arts".

This style was at its height in the late 1920s and 1930s, a period of great economic and political upheaval characterised by periods of economic depression but with some increase in economic fortunes during the mid-1930s. The onset of the Second World War at the end of the decade ultimately formed a watershed in economic, social and architectural terms. McKean (1987: 43) describes the 1930s, stating: "Its sheer variety is typical of a period of transition, for the period was the knuckle between two different worlds – the Edwardian pre-1914, and the Elizabethan, post 1952."

Like the Victorian period, the availability of new materials was highly influential in design. Materials like chrome, vitrolite (an opaque, coloured glass), plywood and concrete allowed new types of building styles to evolve. There were also other technological improvements such as ready availability of electricity, central heating and use of new designs such as flat roofs for buildings.

> In Britain in the 1920s, as elsewhere, the new style was a favourite for building types that had no tradition behind them: garages, power-stations, airport buildings, cinemas and swimming pools. All these tried to be self-consciously modern, and it was not always easy to separate elements derived from the early International Style in the Continent from Art deco in the strict sense.
>
> (Duncan, 1988)

Much of the inspiration in the 1920s arose from the Exposition Internationale des Arts-Decoratifs et Industriels Modernes held in Paris 1924–1925. Curl (1999) states that the 12-volume official publication from the exhibition "disseminated the elements of a style derived from the more severe geometrical patterns evolved as a reaction to Art Nouveau".

Other inspirations included the discovery of Tutankhamen's tomb in Egypt in 1922, which encouraged the adoption of Egyptian motifs and colours. Decoration tended to be geometric in inspiration and in many

26.7   Art Deco façade, Norwich Art School, Norwich

buildings there was a lack of ornamentation, the sleek and decorative-less appearance being the most favoured through the use of smooth materials like concrete, faience and steel.

Buildings in this style included cinemas, hotels such as the Strand Palace, Battersea Power Station and the Hoover and Firestone factories in West London. However, it also inspired interiors; other aspects associated with this time are Tiffany lamps (Louis Comfort Tiffany, 1865–1933) and sumptuous, high-quality interiors like that of the Rogano restaurant in Royal Exchange Square, Glasgow (1935). Increases in car ownership meant that garages were being constructed for the first time.

Cinemas became immensely popular during the 1930s following the advent of synchronised sound in films and the production of numerous films from the American Hollywood studios. As television remained unavailable to the majority of the population, cinemas offered entertainment, and as a result they were constructed across the country and even in quite small towns like Peebles and Crieff in Scotland. Many were daring and modern in their design, adopting materials like faience, chrome and the streamlined designs of the Art Deco period.

26.8   Rogano restaurant, Royal Exchange Square, Glasgow

26.9   Art Deco cinema, Perth, 1933

Designs during the 1930s became very streamlined and modern in appearance, inspired by ideas of aerodynamics and speed. Shopfronts, for example, became flush and sleek with use of smooth, shiny modern materials, epitomising the desire for no ornamentation but still retaining architecture which made a statement in the townscape. Frontages to garages, cinemas and shops tended to be stepped, with detailing inspired by jazz or streamlined themes.

Housing in the inter-war period was different from the pre-war years. Low-rise flats with flat roofs and endless estates of bungalows were common. Styles could be very modern, this being square, boxy and flat-roofed. Windows were typically steel framed and square, or with rounded corners. However, suburban dwellings typically mimicked historic styles with mock-Tudor half-timbering and bay windows with a tile-hung finish. English country styles spread north into Scotland and can be seen, for example, in the small Perthshire village of Forteviot. Haynes (2000: 59) describes this as a "fascinating experiment in rural regeneration, planned before the First World War, by the whisky magnate, John Alexander Dewar". The low buildings with clay-tiled roofs are more typically English Garden City than Scottish in inspiration.

Local authority dwellings were built in greater numbers for the first time. These housing schemes tended to have hipped slated or tiled roofs and although the gardens were small, larger open and green spaces were incorporated as part of the overall design of the schemes. Although perhaps utilitarian by the standards of other houses, the quality of inter-war council housing was still superior to the post-war houses erected for the same purpose.

26.10   Forteviot, Perthshire

Conservation attitudes to early twentieth-century buildings

The fact that until recently, post-1918 buildings were not regarded as worthy of listing indicates the lack of respect for many of the buildings erected in the twentieth century. Although the Edwardian period did receive some recognition, it has been a battle to achieve the same status for inter-war construction, despite the highly innovative nature of this period when architects encouraged new and daring designs to be adopted.

However, the role of groups like the Twentieth Century Society has raised the profile of important twentieth-century buildings and many are now included in the statutory lists.

Chapter **27**

Post-war period

The last 60 years in the United Kingdom have witnessed some of the most dramatic changes and developments in towns and cities. There has been increasing pressure caused by changes in social structure and particularly by the increased use of the car, which has altered the way in which people live and work.

The immediate post-war period was one of rebuilding war-damaged cities like Coventry, Liverpool, London and Glasgow. There was also a tremendous pressure to accommodate people who had been displaced as a result of this war damage. This resulted in the erection of many temporary houses of non-traditional construction which were meant to last for a short period of time. The reality is that many are still in use today.

The pressure to house people resulted in a number of solutions:

- Temporary houses of non-traditional construction such as "Dorran" bungalows or "pre-fab" housing such as those offered by Arcon in England.
- New towns built as "utopian" ideals away from the poor housing of cities. For example, Milton Keynes, East Kilbryde and Cumbernauld.
- Pressure for land meant a solution was high-rise flats, some on land where slums like the notorious Gorbals in Glasgow had previously been located.
- Large-scale council housing schemes were initiated across the country. These were of varying quality and houses tended to be of similar types and were utilitarian in design.

However, in terms of an architectural and social experiment many of these well-intentioned schemes were not successful. Although people were now housed in a far higher standard of accommodation, the splitting-up of communities and the resettling of people away from

27.1   High-rise flats, Glasgow

their original locations proved to be a failure. The harsh urban environments, often with a lack of social amenities and open space resulted in the slums of the later twentieth century.

Commercial building was also brutal in nature during the 1960s. Towns witnessed extensive rebuilding and redevelopment, often whole-scale and savage in its approach with little regard for the built heritage or what it represented. Attempts to accommodate increasing car ownership also meant that new roads and motorways were now being included in town layouts. In cities like Glasgow these cut through much of the existing urban fabric to create motorways which divided previously adjacent communities. While these roads allowed movement around the city, they split communities and destroyed many historically significant buildings.

Concrete became the most popular building material due to being relatively cheap and very flexible in the type of structures it could be used for. Combined with steel it made for new buildings which were often dramatic, and in some cases brutal, in design. Uncompromising, they sat next to their older Edwardian and Victorian neighbours with some difficulty.

The trend continued into the 1970s, although the Civic Trust and the popular conservation movement had been mobilised by this time to try to prevent some of the extensive destruction of older buildings that had been occurring. The statutory listing of buildings had been underway for a number of years and this was at least offering protection to some of the older structures.

27.2   Local authority utilitarian housing

By the 1980s there was an economic boom, despite high interest rates, and office and retail markets expanded. The Thatcher government also introduced the Right to Buy scheme for council tenants which had a dramatic impact on the council schemes by encouraging not only private ownership but also for them then to improve their properties. This undoubtedly raised the standard of some housing schemes as the new owners sought to invest in their houses. However, the housing bubble had burst by the early-to-mid-1990s and negative equity prevailed for many householders, freezing the market.

A more economically stable period followed from the mid-1990s. There was a return during this time to historical styles, shown in major house-builders Barratt and Cala Homes using Neo-Tudor designs and thatched roofs. This takes the erosion of local identity to extremes where historic, sometimes vernacular, styles are used regardless of their location to introduce an idea of a former time-period to potential purchasers. However, the majority of modern housing is much more standard in approach and indeed, in the same way that the increasing standardisation of materials such as slate and brick during the nineteenth century eroded local identity, now identical house styles may be found in the southeast of England and the north of Scotland with no reference at all to local culture, materials or the surrounding environment.

There was an increasing problem faced in the redundancy of buildings, some of them large and problematic with no obvious

27.3   A drawing of the London Eye

alternative use, including coal mines, industrial mills and churches. As congregation numbers dwindled churches were faced with closure, particularly in rural areas but also some in urban areas. Other buildings associated with industrial processes like New Lanark mills, jute mills in Dundee and mines in the northeast of England have all been made redundant as Britain's heavy industries have declined. Finding a sustainable and viable re-use that is also sensitive to the building is a challenge.

The last 20 years have also seen some radical new buildings and structures being erected, ones which may become icons and listed buildings in the future – the controversial Scottish Parliament by Spanish architect Enrique Miralles, the "Armadillo" in Glasgow, the London Eye and the O_2 Dome. These have utilised new construction techniques and materials in the same way that the Victorians and Edwardians exploited their available materials and technology. However, often costly and lengthy projects, they do not always find favour with either professionals or the public and it may only be in years to come that their architectural contribution is appreciated. The controversial building of the glass pyramids next to the Louvre in Paris in 1989 exemplify the fact that such buildings, while brave in their design, also court controversy and it may take a while before they are accepted.

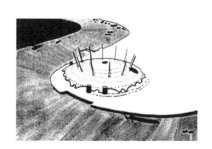

27.4   A drawing of the O_2 Dome   **27.5**   The Louvre, Paris

Chapter **28**

Conclusions

This part has examined the special influence of the Italian Renaissance on the development of building styles in Britain. The Renaissance represented a new beginning in architecture, a "rebirth", and although Gothic had dominated ecclesiastical architecture in Britain, the influence of the Italian Renaissance began to emerge and the Gothic was pushed aside. The eighteenth century witnessed a period of improvement in prosperity in both rural and urban areas with Classical styles being adopted, reflecting the ancient cultures of Greece and Rome. During the Victorian era the boundaries of architectural styles became less clear as Gothic and Classical vied for prime position. The result was often complex and exuberant and distinctly Victorian. To some extent the revival theme was continued into the Edwardian period but, like the nineteenth century, was also influenced by the availability of new materials as well as new styles such as Art Nouveau. During the inter-war period the influence of jazz moderne and Art Deco themes left a legacy of daring and sleek designs, while post-war Britain has struggled to develop an architectural style of its own, although some of the new architecture typified by large public buildings such as the O_2 Dome and Scottish Parliament may prove to be iconic in future years.

Part 4
Vernacular architecture

Part **4**

Introduction

Vernacular buildings span many centuries and do not fit neatly into any particular chronological time period. They therefore warrant consideration in their own right. This part is largely concerned with domestic vernacular architecture from the medieval period until the nineteenth century, although mills, factories and farm buildings also fall into this category and are briefly considered. Given the strong agrarian history of Britain, the comprehension of rural vernacular buildings is an important aspect of architectural history.

It is important at the outset to examine what is meant by vernacular architecture. Curl (1999) defines vernacular buildings as: "Unpretentious, simple, indigenous, traditional structures made of local materials and following well tried forms and types, normally considered in three categories: agricultural, domestic and industrial."

Rice (2006) suggests that styles "developed as one building imitated its predecessor, making use of the most readily available materials … and using them in a fashion unique to that place". The important words are "building imitated its predecessor", as this indicates a succession of buildings continually influencing subsequent buildings throughout history. Buildings develop for many reasons but history and experience are the primary influencing factors.

Brunskill (1981), a leading authority on vernacular architecture, offers the following definition: "A deliberately permanent building which is traditional rather than academic in its inspiration, which provides for the simple activities of ordinary people, their forms and their simple industrial enterprises. It is strongly related to place."

In his book, *Vernacular Architecture*, Brunskill identifies when small utility buildings cease to become vernacular and instead morph into buildings created for specific and identifiable purposes. The small country cottage is certainly vernacular in scale, intent and purpose as

4.01   A vernacular canal wharf building at Froghall, Staffordshire

4.02   A non-vernacular Palladian-influenced silk mill in Leek, Staffordshire

is the larger urban house of a merchant during the sixteenth or seventeenth centuries. However, the large country house or mansion probably is not vernacular and falls into the category of grand and glorious. But, in addition we have identified that agricultural buildings are of the vernacular genre as are small-scale industrial buildings that serendipitously were developed or enlarged to satisfy a local process. Once that process becomes large in scale to satisfy an increasing demand for the materials and goods produced, the buildings in which the process is housed takes on an architectural style and character reflective of the industrialisation of the process. So, a small watermill constructed to satisfy a local need is vernacular, but a large-scale, purpose-built cotton mill may not be.

Although it may be a generalisation, the majority of vernacular buildings were not constructed with a design philosophy that is anything other than utilitarian. Once a building becomes specifically designed for a particular purpose, is large in scale, may use non-local materials and is specifically designed to look good as an architectural statement it probably falls out of the vernacular definition.

So we have a consensus that vernacular buildings are small in scale, unpretentious, make use of locally available materials and respond to local conditions and need in a simple and utilitarian way. However, such a description would almost seem to denigrate what are extremely important buildings in any rural or urban landscape. These buildings are special because they fit in so well with their surroundings; they fit because they were derived from the local landscape through use of local stone, slate and other materials and are, in Brunskill's words, "strongly related to place". They therefore feel like they belong and form a natural part of the landscape. Their contribution to the landscape of any country should not be underestimated.

Chapter **29**

Geography and economy

These definitions indicate that vernacular buildings are highly responsive geographically in terms of the materials from which they are constructed and in respect of their design. Materials were locally sourced and designs were inspired by those materials. For example, the Hebridean blackhouses used driftwood from shipwrecks or whale bones for their roof construction due to their strong connection with the sea (Walker and McGregor, 1996b). Styles therefore vary considerably not only across the United Kingdom but within much more limited regional districts.

In Part 1, we discussed the elements of Banister Fletcher's *Tree of Architecture* and considered the various influences on architecture. Geography and geology are clearly of prime importance. Within England, the limestone belt strongly dictates the nature of stone architecture within it. In other parts, such as the southwest of England and East Anglia, use of earth construction is common. In Scotland, there are considerable variations between the numerous islands and across the mainland. Elements of isolation together with particular trading connections all played a role in defining the buildings and architectural styles that emerged.

The economic fortune of these districts is also reflected in their vernacular architecture. For example, the longhouses of the Scottish Highlands are considerably more humble than some of the contemporary country cottages found in parts of England. Fluctuations in the local economy could have a significant impact on the building styles and whether buildings were erected or extended. For example, the changes in England's wool trade had a considerable impact on Lavenham in Suffolk. Lavenham became very prosperous through the wool trade and by the early 1500s was one of the richest towns in the country. However, an influx of Flemish weavers and heavy taxes caused a subsequent decline in the town's fortunes. The buildings that stand today are a reflection of both that economically successful era, together with a subsequent period of stagnation and absence of intervention. This later period of inactivity helped to preserve the village from change for a sustained period of time. Vernacular buildings are therefore strongly connected to the success of the local agrarian and wider regional or national economy.

Taking into account economics, geography and geology, it is clear that the study of vernacular buildings can be complex. An understanding of particular local conditions is a prerequisite to understanding the buildings to be found there.

Chapter **30**

Early vernacular buildings

Very little survives of vernacular buildings under the feudal system of the early Middle Ages when the common man was subject to his master's will and direction. He was not capable of acting independently and was therefore not able to create buildings other than for simple shelter using whatever was available and best suited the local climatic conditions. None have survived in recognisable form so we must rely on archaeological investigation and, possibly, speculation as to their exact form. Almost certainly they would have been simple structures with the minimum of comfort. It is also important to appreciate that there would have been a difference between rural life and what might be termed existence within settlements centred on castles and the defended houses of the nobility.

It is not until the end of the Middle Ages that surviving vernacular structures provide us with examples that truly reflect what life might have been like for common people in the late fifteenth century. These remaining buildings provide us with evidence of plan, fenestration and decoration as well as living conditions, constructional methodology and use of materials, although they may have been altered significantly over their long lifetime.

Early aisled houses and hall houses of the late fifteenth and early sixteenth centuries were simple structures with possibly a single or even double solar to provide sleeping spaces at each end, located above the ground-floor services level. This is typified by the Wealden houses of the southeast of England. Such houses would have been timber framed with either a thatched or tiled roof. There were no chimneys and only a hole in the roof or gable to let the smoke out from the fire pit which was usually located in the centre of the main, sometimes two-storey hall.

In other parts of the country, such as in Cumbria and Scotland, Wales and the West Country, these houses are more likely to have been of stone wall construction as stone was readily available. In the rural areas this will have taken the form of rough field boulders rather than dressed stone. Where available, lime would have been used, but in clay areas, this would often have been dug out of the area close to the building and used as a mortar, sometimes mixed with straw and dung.

Roof construction would have been in timber, perhaps a cruck frame under a straw thatch, stone slab or turf roof weighted by large stones tied together to resist the damaging effects of high winds in exposed areas. Heather was also used as a roofing material in highland areas where it was in plentiful supply. However, the scarcity of timber in certain localities meant that this was the most valuable part of the building and was even removed by tenants when they moved to a new location. In Scotland a "hingin' lum" of clay and wicker was sometimes formed within the roof-space as a canopy to help draw the smoke up and out of the room, but in the blackhouses of the highlands and islands the houses were filled with smoke and the lack of natural light must have made them highly unpleasant to live in. Glass would have been prohibitively expensive for all but the highest class of buildings and instead wooden shutters (known as "hurdles" in Scotland) would have been used.

Pre-dating these cottar dwellings, one of the most interesting early houses found in Scotland and Ireland are loch dwellings known as crannogs. These are round houses built out into the lochs, usually on timber-piled foundations. They were used from 5000 BC but were still occupied in Scotland until the seventeenth century. These highly unusual dwellings housed families in their thatched and timber construction and a recreated one has been built on Loch Tay, Perthshire. The original construction methods were discovered using underwater archaeology and this detailed investigation allowed the accurate reconstruction to be undertaken.

Strong regional differences are typified by the longhouses found in the highlands and islands of Scotland. Known as blackhouses in the Hebrides, these cottages once prevailed throughout this region, but their construction declined significantly in the twentieth century although they were still being constructed up to the end of the nineteenth century. Walker and McGregor (1996) describe them as:

> long, low, narrow-bodied, chimney-less byre-dwellings where the human occupants and the cattle shared the same door and the same internal volume. It also represents one of the final stages in the development of an ancient building type built to entirely different principles to those in general use today.

30.1   Recreated crannog, Loch Tay, Perthshire

30.2   Camserney Longhouse, Perthshire

30.3 Elm Hill, Norwich – timber-framed buildings dating from the fifteenth and sixteenth centuries

The walls were of locally sourced field boulders with an earth core. The roofs were thatched with turf and heather and window openings were very small. They typify the definitions of vernacular outlined above with their use of entirely locally sourced materials, utilitarian and strongly associated with the place where they exist. However, their apparently simple construction belies a much more complex approach to architecture than may be thought by first impressions. The sophistication of construction techniques is an aspect of vernacular buildings that is often misunderstood.

Although there were considerable geographical variations, by the sixteenth century vernacular buildings were becoming more prestigious. The rural areas would have maintained their traditions and perhaps simpler cottages, but in the wealthier towns, the urban landscape began to reflect the increasing wealth of their citizens. Small civic buildings such as those created by the emerging guilds and merchants demonstrate the importance of these bodies and are also indicative of the developing wealth of groups in society other than the nobility and royalty. They also demonstrate the diminution of the Church as a formerly dominant power in medieval society.

Chapter **31**

Vernacular houses

Houses were built to meet the various requirements of the occupants in terms of protection from the weather, a place to sleep and cooking facilities, usually an open fire. Larger houses would also offer a place for gatherings. The early and simpler houses had less division of space and indeed, in some rural buildings such as longhouses, there was little separation between the space for people and that for the animals. However, where the economic circumstances permitted, there were opportunities to enlarge and diversify space and purpose. An examination of the layout plans of houses can therefore tell us about the history of a house, its age and purpose. Brunskill (1981) identifies five main types of plans for vernacular houses as H-shaped hall houses, Wealden houses, cross-passage, baffle entry and double-pile houses:

H-shaped hall houses or wing houses: medieval manor houses during the mid-fifteenth century had a central hall with cross-wings at either end (to form an "H"). At one end of the hall there was a screen passage and at the other a dais or platform giving a raised area for a high table.

Wealden houses: these were first recognised in the Kent and Sussex areas and have a central hall with service rooms on one side and private rooms on the other. Typically, the Wealden house would have also accommodated an open first-floor shelf or solar to provide separate sleeping accommodation. Initially these would have had an open fire in the main hall, but over time brick chimneys were introduced. Generally smaller than the hall houses, they have distinctive jettying at the first-floor level and often steeply pitched, hipped roofs. The Wealden houses were essentially yeoman farmers' houses and were an important reflection of status, prosperity and quality.

Cross-passage houses: this is a common arrangement and houses like this were found from medieval times until the seventeenth century. The fireplace backs onto the cross passage which typically runs from the front to the back of the house floor plan.

Baffle entry houses: these houses have a small entrance lobby and are characterised by large chimney stacks. These types are commonly found in East Anglia.

Double pile small houses: this arrangement is for a house which is two rooms deep and usually with the fireplaces located in the side walls of the building.

The development of the brick chimney marked an important progression in house plan forms. They became a status symbol and highly decorative chimney stacks are a reflection of this. These enclosed fires allowed the whole house to be heated, not just the main hall, and there could then be a clearer division of household purposes such as working kitchen areas and bedrooms. The fireplaces allowed houses to become more compartmentalised and floors to be inserted.

Over time these houses have been altered and their original plan form may therefore not be immediately obvious. A Wealden house tends to be obvious because of the external jettying but even then, alterations may have taken place which can disguise these original features. It is therefore important to understand these plan types together with their methods of construction in order that their historical significance may be appreciated.

Timber-framed houses and methods of construction

The majority of early civilisations used timber to facilitate their built environment. It has been used not only for permanent or semi-permanent buildings but was also used to facilitate and assist in the construction of more permanent structures, either as scaffolding, temporary support, or centering for arcuated forms of construction. It is used as both structure and cladding and, despite its susceptibility to damage by fire, can, if used with an appropriate section, remain stable during and after fire due to its natural ability to form a resistant barrier through carbonation. Timber is vulnerable to insect damage but certain timbers have natural resistance – for example, teak can resist termite damage. It can even resist water damage and deterioration if oxygen is excluded or limited. The sixteenth-century *Mary Rose*, the oak warship raised from the Solent is a classic example of this process. So, we have a continuum of use and incorporation of timber within our built environment from pre-history to the present day, its only probable limitation of use being what is or is not acceptable in fashion terms.

The Church had a declining influence on society from about the fourteenth century which culminated in dramatic form in England in the mid-sixteenth century in the Reformation and Dissolution of the Monasteries under Henry VIII. The use of timber constructional forms for vernacular buildings subsequently declined during the late sixteenth and early seventeenth centuries as timber-framed buildings became unfashionable and brick and stone construction became much more acceptable. During the seventeenth century many cities were ravaged by fire and new regulations to prevent or mitigate fire damage were introduced. This rendered timber buildings with thatched roofs as non-compliant. Alternative, less vulnerable forms of construction therefore rose to ascendancy. This decline was also encouraged by the increasing shortage of suitable timber during the late sixteenth and early seventeenth centuries, probably due to the rapid rate of use of timber for building and ship construction. However, some cities continued a timber tradition into the seventeenth and early eighteenth centuries including London, Manchester, Chester and Shrewsbury (Brown, 1982).

The use of timber framing was widespread but varied locally in terms of design. Timber-frame methodology may be separated into two or three main forms: the cruck frame, the post and truss form and box frame. It might be correct to link the cruck truss and the post and truss methods under one generic grouping – the cross truss. The main construction types are outlined as follows:

Cruck frames: there are different types of cruck construction but the principle is similar to the use of a triangular frame, usually as a 'blade' or curved timber. Pairs of timbers, sometimes large and rustic in profile, are spaced at intervals with purlins and wall plates to offer lateral restraint and roof support. Types include a base-cruck where the cruck blades are situated close to the ground and upper crucks where they are positioned on the head of the masonry ground-floor walls of the building. Additional variations are: middle cruck where the frame sits at a half-storey height of the ground floor, usually on half-storey ground-floor masonry walls; and jointed cruck where the frame is made up of two sections with the lower section forming the wall framing at ground floor. Cruck construction may also be used in combination with box-frame wall construction. These types of construction are particularly found in Scotland, the Midlands, northern England, central Wales and especially in Hereford. Their use in Scotland is associated with croft buildings such as Camserney Longhouse in Perthshire (see Figure 30.2).

Box frames: this uses horizontal and vertical timbers to create a box. The walls are then infilled with material such as wattle and daub or brickwork. These exhibit regional variations with close studding found in the east of England and square panels found in the west. The close studding is characteristically Elizabethan and was used for display purposes, creating highly decorative patterns. These are found largely in the south and east of England.

Post and truss: the principle here is similar to cruck where, unlike the box frame, the weight is taken by the roof trusses into the vertical posts. Curved wind braces are typically found under the rafters and horizontal purlins may be evident at the gables.

These forms of construction vary in terms of the way they distribute their loads and also in the way in which they are initially erected. Cruck framing distributes the vertical loads via crescent-shaped frames to foundations provided by perimeter plinth walls or stonebases. In contrast, box frames distribute loads via a network of interdependent and interlocking frames, evenly distributing wall and roof loads to the footing or foundations. Invariably cruck frames and some post and truss frames were pre-assembled flat on the ground and then "erected" and "tied" together by ridge beam, purlins and wall plates. Box-framed construction (for the most part) and some post and truss frames were erected as a developing form from vertically erected posts, temporarily supported, with infilling members fitted to the posts as the construction extended.

It is evident that there is particular regional distinctiveness associated with timber-frame construction, although the pattern is complex. There are local variations in style, approach and in jointing details. For example, the cruck frame is limited from Oxfordshire in the southwest to Yorkshire in the northeast, Flintshire in the west and to Lincolnshire in the east. However, the use of cruck frames does not appear to have been adopted in the southeast where post and truss and box framing is more prevalent. Box framing was adopted in most areas whereas cruck framing had a more geographically limited area of use – mostly, though not exclusively, to the west of the limestone belt running from the Severn estuary to the Wash. Within the same approximate geographical area post and truss forms were also adopted (at a later date). Using a similar constructional approach as cruck frames, the post and truss forms adopted independent framing at regular centres (usually about six feet apart) and linked together by purlins and wall plates in the same way as cruck construction.

As well as adopting a structural role, timber framing also had a decorative purpose although this embellishment was more popular in some regions than in others. Close studding, with its numerous vertical timbers, was a later development dating to the late sixteenth and early seventeenth centuries. It is characteristically Elizabethan in style and is found especially in the south and east although it can be found in other areas (see Figures 31.1 and 31.2). However, it waned in

31.1   Close studding in a building in Leek, Staffordshire

popularity once oak timber became scarce during the late sixteenth century. The exuberance of the Tudor period therefore gave way to the subsequent more austere styles due to the lack of availability of suitable timber. Decorated timber framing is particularly prevalent in the Midlands, north and west England from the seventeenth century. Decorative details included trefoils and stars and can be seen in towns like Ludlow in Shropshire and in Cheshire (see Figures 31.3 and 31.4). These regional variations, while indicative of localised differences, should not be assumed to be hard and fast rules associated with location as the exception can prove an expert wrong, but they are typical of the regional variations that help to identify stylistic influences associated with certain areas.

In addition to using the timber itself in a decorative manner, the infill panels between the timbers also offered an opportunity for changing the aesthetic appearance of the building. Infilling of the frames typically adopted timber panelling or wattle and daub. Brick nogging was used from the seventeenth century onwards as infill between the panels, replacing the formerly used wattle and daub as brick became a more fashionable material. Brick nogging or infill could be in various patterns but was often herringbone in design (see Figure 31.5). In Norfolk, Suffolk and north Essex timber-framed houses were invariably rendered, sometimes with elaborate pargetting or decorative plasterwork. This covered up the timber

31.2   Studding at less close centres in a much-modified former timber-frame building in Norfolk

31.3   Little Moreton Hall, Cheshire is an example of a heavily decorated timber box-framed building (Tudor)

31.4   Decoratively framed Ludlow buildings

framing as did mathematical tiles. These tiles could be used to hide the original framing structure and make the building appear as though it was constructed of brick at a time when brick was more fashionable. Mathematical tiles are clay, geometric tiles designed to represent brickwork, hung on battens and fixed to earlier, 'less impressive' timber structures. They were developed and widely used in the late seventeenth century in certain areas to raise the apparent status of the existing timber buildings.

31.5   Close studding with herringbone brickwork as panel infilling – the brick infilling is likely to be a replacement for previous wattle and daub (Norwich)

31.6   Rendered timber-frame building in Norwich with render effect – a form of pargetting to represent rusticated ashlar stonework, in this case rather crudely done

31.7 The building on the extreme right is an original timber-frame house to which, on the first and second floors, mathematical tiles have been added as cladding

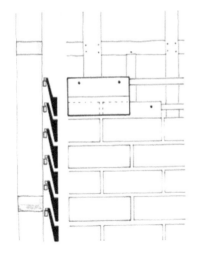

31.8 A drawing of mathematical tiles

Jettying

In urban situations land was valuable and owners tried to achieve as much floor space as possible on limited plots by overhanging the storeys above ground level. This method of construction is known as jettying. The weight of the overhanging floor, acting as a cantilever, was counterbalanced by the weight and effect of the floor within. So, in theory, successive upper floors could be well outside the ground-floor plan area. Although the jetty probably originated in urban situations, it was perpetuated in rural locations where space was less of a premium thus indicating a 'fashion' response not strictly generated by need. This method of construction may therefore be associated with status and it was a method widely adopted in the sixteenth century.

At the corners of the building and supporting the wall construction at the corners is a floor beam termed a 'dragon' beam. This radiates diagonally at the corners and facilitates the structure where the floor joists need to 'change direction' but retains the counterbalance weight of the floor where the change in span direction would limit applied loads to resist the cantilever. In Figure 31.11 it can be seen that the

31.9   A jettied first floor with a cantilever of about 600 mm

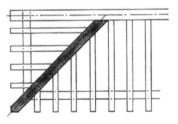

31.10   A dragon beam

31.11   A thickened corner post to support a dragon beam at the first floor – the console bracket at the first-floor ceiling level to support the jetty dragon over

corner post has been thickened by a console bracket to assist in support of the dragon. The beam at the jetty edge and supporting the floor above is called a "bressumer" beam. A beam supporting the inboard edge of a jettied floor over and forming the stop to the wall frame below is called a "summer" beam or "jetty plate".

Chapter **32**

Vernacular farm buildings

Differences in farming practices meant that specialist buildings developed over time. Some of these were regionally distinctive depending on particular local circumstances of economy, topography and available local materials. These buildings were all erected to serve particular functions and are therefore designed with functionality in mind. However, because farms vary so much in their functions, the building types reflect this and are numerous in terms of style and design. Some of the most important are listed below:

Tithe barns: These are the great medieval barns where a tithe or tenth of the profit from the land was levied by the Church in order to pay for the upkeep of the local parish. The largest monastic barn is believed to have been at Cholsey in Berkshire and was 92.4 metres long by 16.5 metres wide. It was demolished in 1815 (Brunskill, 1999: 37). Walls were usually, but not always, of stone with either a thatched or stone roof. Openings were limited to ventilation and access doors.

Aisled barns: These great barns have a central nave with aisles at the sides and tall, steeply sloping roofs, again with few openings other than for access or ventilation.

Oast houses: These are found in Kent and were used for the drying of hops. The characteristic conical roofs are located above the kiln used to dry the hops.

Horse-engine houses: These buildings are usually single storey and either round, apsidal or multi-sided. Inside they used mechanical threshing machines driven by horses. They may be of stone, brick or timber construction, with varying openings. Roofs may be thatch, tile or slate.

32.1   Waxham Great Barn, Norfolk

32.2   Vernacular barn, Normandy

The above list of types is by no means exhaustive and other forms include stables, smithies, dovecots, granaries and other forms of barns and storage buildings. All are typified by an architectural response to provide simple buildings whose primary purpose is to fulfil a utility and unpretentious working function.

Chapter **33**

Vernacular industrial buildings

In addition to farm and domestic use, buildings were erected locally to serve various industrial purposes. Like farm buildings, their function, and therefore design, varies considerably. The most important types are discussed here although, like farm buildings, there are so many different types that this is a specialist subject in its own right.

Watermills

Watermills have been in use since Roman times and certainly there are remains associated with sites such as Hadrian's Wall. Like other vernacular buildings, there are considerable regional variations in types and adaptations. For example, in certain parts of the country, notably Orkney, Shetland and the Outer Hebrides, horizontal Norse mills are found, indicating the Viking connections of these islands. In Shetland, the mills are small low stone buildings located on burns, making the most of the local topography and watercourses.

Watermills were largely used for the milling of grain and were strongly associated with monastic settlements in order to provide for the religious communities located there. A famous example is James Brindley's corn-grinding watermill at Leek in Staffordshire. James Brindley (1716–1772) was originally a miller but is famous as a pioneer of the Industrial Revolution canal-building programme. His study of water management at his mill sites germinated an interest in water engineering and he worked as both a millwright and later as a canal engineer. The canal system, developed during the Industrial Revolution, generated a whole architectural and engineering genre based on its requirements.

Watermills, in addition to milling grain, may also be used to grind other materials such as stone or flint to support associated and localised industrial processes. Figure 33.2 shows a flint mill at Cheddleton in Staffordshire which supplied crushed and ground flint for use in the production of pottery in Stoke-on-Trent.

33.1   Locks and bridges of the Trent Mersey canal at Stone in Staffordshire

33.2   Cheddleton flint mill, Staffordshire

Included in this water-driven technology are waterwheels. These provided power to a changing industrial landscape during the late eighteenth and early nineteenth centuries. They were minor buildings in scale but they had a significant impact on the industrial processes that were developing at that time. Waterwheels transformed the energy in fast-running water into power for belt-driven machinery. This transformed the cloth-making process from a fundamental cottage-based industry to one that developed during the Industrial Revolution and facilitated the immense changes that typified that period. It was of course later usurped by steam technology which also facilitated great changes to the mining industry. Pumping out deep ore seams and allowing, in the tin-mining industry for example, shafts to be excavated well out to sea in tunnels that would otherwise have been impossible to mine due to ingress of water. All these early Industrial Revolution buildings are of a vernacular type but are very important to conserve as evidence of our industrial heritage.

33.3   A drawing of a tin mine pump house in Cornwall

Windmills and windpumps

These were used to power milling equipment and also for drainage. Styles and materials vary locally and include tower mills, usually of circular brick construction, and post mills usually of timber construction. Tower mills are mostly, though not exclusively, found in East Anglia and post mills elsewhere.

33.4   Tower mill at Sutton, Norfolk   **33.5**   Windpump near Stalham, Norfolk   **33.6**   Thurne Dyke windpump, Norfolk

According to Watts (2000), windmills were found in Britain from around the end of the twelfth century and they spread during the thirteenth century, especially in the flat lands of East Anglia and the Fens where water power was not as effective. The early timber post mills were vulnerable to storm damage and so stone or brick towers, which were more resilient, were built. The earliest stone tower which survives is believed to be at Burton Dasset in Warwickshire.

Brunskill (1999: 53) states that although some farms in the eighteenth century used water and wind power for operating threshing machines, these methods were not regarded as reliable enough. Horse-engine houses therefore became the preferred method of mechanisation by the end of the eighteenth century.

33.7   Post mill at Upton, Norfolk

Chapter **34**

The conservation of vernacular buildings

There are a number of issues with vernacular buildings:

- lack of understanding of their methods of construction due to lack of research or poor survival rates;
- lack of suitable skills for repairs;
- lack of suitable re-use alternatives, particularly for industrial buildings.

Vernacular buildings are complex structures and the reduction in local and specialised construction skills has particular implications for vernacular buildings. While humble and apparently simple in their construction, they make a significant contribution to the rural landscape and to those of small villages and towns. Inappropriate repairs and alterations can seriously compromise the integrity of these buildings.

With some types of buildings there is pressure to convert to residential use and this may be particularly unsuitable for farm and industrial structures. In these buildings there tend to be few openings so the need to meet modern domestic requirements can be detrimental to these buildings.

Vernacular buildings, while simple in their source of materials, are often extremely complex in their construction and frequently their exact nature is misunderstood. This lack of appreciation for the mechanics that underpin these buildings can lead to inappropriate approaches to repair and maintenance leading to deterioration of these structures. Research at this level of the built environment is worthy of study in its own right and cannot be covered in sufficient detail in this volume, but it is still very much the responsibility of the conservation movement to maintain and protect vernacular structures.

In addition, to understand the former uses of what may have become redundant buildings the conservation practitioner needs to realise that

34.1 A converted farm building

these structures may have become equally important to new occupiers – indigenous wildlife. Part of the process of evaluation prior to intervention must include an investigation of the forms of wildlife that may now inhabit redundant structures. The absence of human use and occupation has provided some of our rare species with safe and secure habitats, and interventions – either in the form of repair, maintenance or conversion – may be placing a severe risk on these wildlife alternative users. It is therefore an essential element of intervention planning to assess and accommodate the use of redundant structures by wildlife. Bats and barn owls are particularly at risk when conversion of redundant farm buildings is being considered and the Wildlife and Countryside Act 1981 has been placed on the statute book to ensure protection of our rare species.

Chapter **35**

Conclusions

Vernacular buildings are extremely varied in their materials, styles and age. They are essentially utilitarian in nature, built for a specific purpose, and reflect locally available materials. They form an intrinsic and important part of the built heritage but a lack of understanding of their importance and methods of construction has meant the loss of many of these buildings. Increasing awareness through the workings of groups like the Scottish Vernacular Buildings Working Group and other similar societies means that the importance of these buildings and the research required into the origins and construction is being raised.

The following books offer the reader a deeper understanding of particular aspects of vernacular architecture:

Clifton-Taylor, A. (1987). *The Pattern of English Building*. London, Faber & Faber.

Mercer, E. (1979). *English Vernacular Houses*. London, RCHM.

Brunskill, R. W. (1982). *Illustrated Handbook of Vernacular Architecture*. London, Faber & Faber.

Brown, R. J. (1982). *The English Country Cottage*. London, Robert Hale.

Brown, R. J. (1982). *English Farmhouses*. London, Robert Hale.

Penoyre, J. and J. Penoyre (1978). *Houses in the Landscape: A Regional Study of Vernacular Building Styles in England and Wales*. London, Faber & Faber.

Essex County Council (1994). *Regional Variations in Timber-Framed Buildings in England and Wales. The proceedings of the 1994 Cressing Conference*.

Beaton, E. (1997). *Scotland's Traditional Houses: From Cottage to Tower House*. Edinburgh, The Stationary Office.

Fenton, A. and B. Walker (1981). *The Rural Architecture of Scotland*. Edinburgh, John Donald Publishers Ltd.

Brunskill, R. W. (1981). *Traditional Buildings of Britain*. London, Victor Gollancz.

Naismith, R. J. (1989). *The Story of Scotland's Towns*. N.p, Edinburgh.

Harris, R. (1995). *Discovering Timber Framed Buildings.* Shire Publications, Aylesbury.

Alcock, N. W. with M. W. Barley, P. W. Dixon and R. A. Meeson (1996). *Recording Timber Framed Buildings: An Illustrated Glossary.* Council for British Archaeology, London.

Innocent, C. F. (1999). *The Development of English Building Construction*. Shaftesbury, Donhead.

Stenning, D. F and D. D. Andrews (eds) (2002). *Regional Variations in Timber Framed Buildings in England and Wales down to 1550.* Essex County Council, Chelmsford.

English Heritage (2007). *The Conversion of Traditional Farm Buildings: A Guide to Good Practice.*

Part 5
Materials and performance

Part 5

Introduction

Parts 2, 3 and 4 have examined how architecture has evolved over many centuries and how these changes have been influenced by strongly inter-related factors including climate, geographical location, fashions and trends, human needs and, notably, available local materials. It is essential to understand materials, their geographic spread and how they have been used or even misused if buildings are to be understood. Understanding the performance of building materials and construction methods together with an appreciation of the processes of deterioration over time is a fundamental requirement for any conservation practitioner. The ability to investigate and assess materials, methods and deterioration processes will provide the conservationist with the necessary knowledge that will underpin any intervention works.

This philosophical approach is outlined by Pickard (2000) who states: "Before the design and specification of repairs can be determined it is essential to fully assess the nature and condition of the structure and building materials and the causes of defects to these factors in an historic building." This investigative ability will assist the practitioner in choosing between many, sometimes conflicting, options when planning intervention work. This process is outlined by Bell (2001) who states that the conservation process must encourage the following abilities:

- hypothesise (and test hypothesis) from a clear understanding of the basic principles;
- synthesise the implications of disparate factors; and
- think laterally and creatively in the search for a minimally interventive solution.

An understanding of materials is clearly central to this process. We need to appreciate the limitation of materials, constructional techniques and how these affect the condition and longevity of historic buildings. Conservation response is not simply about preservation, it is a synthesis of many, sometimes disparate, factors. It is about the ability to manage appropriate change informed by detailed knowledge of both condition and cultural significance. It is therefore crucial to know what is happening to a building, how it may be deteriorating and how it will react to change. From this understanding comes the ability to make appropriate decisions.

This part will therefore give an overview of the most important materials used in building construction, how they function and deteriorate and how that potentially impacts on the longevity of the built heritage. This is a vast topic and we can therefore only hope to look generally at these materials and briefly consider how their nature and performance has implications for conservation. The reader is therefore advised to consider, in more detail, many of the more specific texts on particular materials and their properties, where appropriate.

Chapter **36**

Lime and cement

There are two principle forms of binder typically used. In traditional buildings lime is the most commonly used but, more recently, cement has become the favoured material for modern buildings. They differ considerably in the way they set: lime sets by carbonation and cement by hydration. The comparison between these two methods of setting goes a long way to understanding the differences between the two products. A set achieved by carbonation produces a material composite which retains an ability to breathe or be permeable. The resultant set using cement or cement-based products achieves a chemical set by hydration which does not retain permeable properties and, therefore, does not allow the resultant set product to breathe and is impermeable to the passage of moisture.

Historic buildings before the mid-to-late nineteenth century will, invariably, have used a binder based on lime. This section will examine the various types of lime and their particular hydraulic properties, notably that some have a hardening based not only on simple carbonation, but also on an element of hydraulic set. Although lime is basically a carbonation binder it may also have, within a limited range, the ability to set via hydraulic process.

Cement on the other hand, because of its chemical composition of calcium carbonate (chalk and/or limestone) and calcinated clay in higher proportion than simple but hydraulic lime, achieves a chemical and stronger set by hydration or reaction with water, even in the absence of air. It is stronger in compression than lime and is resistant to the passage of moisture. It is ideally suited to contemporary construction methodology and it has allowed the creation of structures that would not have been otherwise possible using lime technology. For example, long spanning bridges and high-rise buildings might not have been possible but for Joseph Aspdin's patent for Portland cement in 1824. However, cement is generally inappropriate for historic structures where lime was used in the original construction.

It is crucial to understand the differences between these two basic binders because they potentially have a fundamental impact upon the historic built environment if either is used inappropriately, such as for repairs or alterations where these were not the material originally used.

Allowing historic buildings to breathe

Prior to the introduction of damp-proofing methods, all buildings were subject to the migration of water from both ground and air. This was either in the form of penetrating damp or condensation, both surface and interstitial (within the thickness of a structure). Moisture would also have migrated from within the structure to the outside, either through seasonal variations in temperature and moisture pressure or simply as a result of the building being used and heated.

Condensation is formed when water within air can no longer be sustained as vapour by the air retaining it, usually due to a drop in temperature or pressure. All air contains water in vapour form, referred to as humidity. Cold dry air will have low moisture content (low humidity) and warm moist air will have high moisture content (high humidity). If the air temperature is lowered the water vapour must condense out to form condensation, either on the surface of colder material or at a point in space where the air temperature drops below its ability to maintain water vapour within it. This transitional temperature is called the dew point and may occur on the surface of a cold material or within the thickness of a structure (interstitially). The humidity level in the United Kingdom will be between 85 per cent and 55 per cent, whereas in the tropics it might vary between 95 per cent and 53 per cent, dependent on season and temperature in both locations.

Historic buildings need to breathe in order to maintain the moisture equilibrium between warm humid air and colder drier air. The movement from warm moist to cold dry air is called moisture movement or migration and can, if unobstructed, move from inside a building to outside. Air will naturally move through a permeable structure to create a balance or equilibrium between the positive moist air and the drier negative air. It is the ability of old buildings to breathe in this way that is one of the fundamental knowledge requirements of any practitioner involved in historic environment intervention work.

If an impervious material, such as an externally applied cement render, blocks this breathing path, the structure may suffer from an excessive build-up of moisture. This may then cause a deterioration of the building fabric. Additionally, the wetter an element of a structure becomes the colder it gets. The result is a vicious cycle of decline due to increased moisture content and a concurrent drop in temperature encouraging further moisture absorption and further temperature drop and so on. Some of this temperature drop, in wet structures or fabric, is due to latent heat loss associated with evaporation of moisture from the fabric or structure. In simple terms, some of the temperature from the building is transmitted to the atmosphere as the moisture contained in the fabric is given off to drier or warmer air surrounding it.

The result of the use of inappropriate, impermeable materials is likely to be an increase in moisture content within a wall element. This increase in moisture content will generate the potential for a build-up of both interstitial and surface condensation and an increase in moisture attracted from the ground by capillary action (rising damp). This may lead to a deterioration of both wall finishes and, potentially, the structure itself. A classic example of this is blown plaster at the base of an external wall where inappropriate gypsum plaster and modern impervious paint has been used in 'renovation' work.

Therefore, understanding the ability of lime to allow moisture movement is one of the fundamental knowledge tenets that all conservation practitioners need to acquire. Old buildings need to be allowed to get wet and then dry out in order to maintain their moisture equilibrium. This reduces any hazardous build-up of moisture that might otherwise create conditions that could lead to deterioration as a result of chemical, biological and insect damage.

In contrast, modern buildings function in a totally different way based on a design philosophy and constructional technology that attempts to exclude water from both the building fabric and the structure. This therefore prevents any detrimental build-up of moisture within the fabric and structure which cannot be easily dispersed by air movement through it. The modern building control requirement to minimise air leakage and thus retain warmed air within the structure typifies this approach to limited permeability building construction technology. So, modern methods of construction use impermeable materials and effectively seal buildings in order to retain valuable heated air and to keep the structure dry by the exclusion of moisture.

This simple difference of construction methodology adopted in old and new buildings must be fully comprehended if inappropriate and damaging intervention is to be avoided.

Use and properties of lime

Early building materials were bound together to produce a structural or enclosing wall in order to support a roof. Earth structures were naturally bound by the use of mud, clay or earth in various forms, such as used for sun-dried mud bricks, cob, pisé de terre and clay lump. Harder material such as stone, fired bricks and tiles required a binding material that was more enduring than mud or clay. Early stone buildings relied on the expertise and skill of the stonemason to ensure that the joints between elements were so closely fitting that such binding materials or mortars were not always necessary. For example, the stonemasons of Ancient Egypt and Greece were so expert at tooling and finishing the stone components that no mortar was used, as indeed was the case in some of the buildings of ancient Mayan and Inca cultures. However, lime as a binder has been used for 7,000 years and really its use only declined after the introduction of Portland cement.

Lime for building is produced by quarrying and crushing limestone (calcium carbonate) or chalk, then calcinating, or burning it within a kiln at temperatures greater than 850°C. This process burns off the carbon gas within the limestone to produce calcium oxide or quicklime. If this product is then mixed with water (slaked) to produce

36.1   Close-jointed stonework at Machu Picchu, Peru

36.2   Stonework detail, Machu Picchu, Peru

slaked lime (calcium hydroxide) it can be used to bind coarse but well-graded sand to produce lime-based mortars and plasters. If left open to the air, the calcium oxide would quickly reabsorb carbon dioxide from the atmosphere to return to its former carbonate state. To prevent this from occurring, the calcium oxide is mixed with water to form calcium hydroxide and, if kept under water, will exclude air and carbon dioxide thus inhibiting setting.

It should be noted that slaking of calcium oxide or quicklime is a dangerous and hazardous operation as the quicklime is caustic and its reaction with water is both violent, heat producing (temperatures of about 300°C are common at slaking stage) and explosive, so great precautions are necessary during this phase of production.

The chemistry of lime and its ability to bind mortars and plasters is demonstrated in simplified form by the well-known "lime cycle".

The process is demonstrated in the production of lime at Froghall in Staffordshire. Limestone was brought down from quarries at Cauldon Lowe in Derbyshire to Froghall and burned to produce quicklime which was distributed to the industrial Potteries and elsewhere via the adjacent Cauldon canal. Local production was served and communications developed to support a local process, so it is important to understand the social and contextual significance and the reasons why certain operations and settlements occurred where they did.

Early use and production of lime would have occurred on the site of construction. The calcium oxide (quicklime) would either be delivered to the site or produced there *in situ* in kilns, and slowly added to water. The lime putty thus produced would have been immediately mixed with coarse sand to form a mortar. This mortar would have been used immediately or stored on site under cover of wet canvas, sacking or tarpaulins in order to reduce air/carbon dioxide absorption

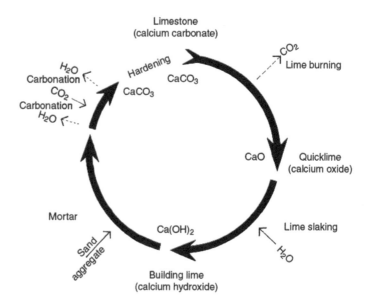

36.3   A diagram of the lime cycle

36.4   Lime kilns at Froghall, Staffordshire

until needed. Stored in this way the mortar can be kept for several days and then "knocked up" by re-working and then used as necessary. Modern lime putties are supplied in plastic buckets with an excess of water so air is constantly excluded from the lime, avoiding re-carbonation until used with sand to produce mortars that set by evaporation of water and absorption of carbon dioxide.

Ideally, slaked lime should be allowed to mature for at least three or four weeks, or preferably longer, to ensure that all the quicklime has reacted with water. Any quicklime in lime putty that has not reacted fully with water remains as quicklime and such particles in materials like mortar or plaster can have an adverse affect on the mortars and plasters produced. Damage can occur, caused by rapid expansion of the unconverted quicklime as it absorbs carbon dioxide from the air. It can also cause lime leaching on the surface of brickwork laid in mortars using inadequately matured lime putty. This is evident as an unsightly white bloom that appears on the brickwork.

Lime mortars require the use of coarse but well-graded sand, as opposed to the finer sands used in cement mortars. The coarser but well-graded sands are rougher than those used in cement mortars, which provides a greater surface area of aggregate for the lime binder to adhere to. Properly mixed lime mortars will produce a very "fatty" or sticky mortar that will adhere to an upturned trowel or plasterer's mortarboard or float, hence the use of lime as a plasticising additive in cement mortars.

Lime sets by absorbing carbon dioxide from the air but it takes a considerable time for enough carbon absorption to occur to achieve a set. However, in wet conditions a set cannot be achieved unless there is clay present as this allows a partial hydraulic set to occur. Although pure lime putties or non-hydraulic limes do not contain any of these contaminants, certain limestones and chalks do contain clay. The varying levels of the silicates and aluminates contained in the clay contaminants will affect the hydraulicity of the lime, setting rates and compressive strengths. These qualities would have been both understood and manipulated by the masons and bricklayers of earlier times, but this skill has been largely lost during the twentieth century as a result of the decline in the use of lime as a binder.

Limes produced from limestones with these different contaminants were identified as feebly, moderately and eminently hydraulic. We now refer to these as NHL 2, NHL 3.5 and NHL 5. The 2, 3.5 and 5 refer to the compressive strengths achieved after 28 days. NHL is simply a shortened version of *N*aturally *H*ydraulic *L*ime. However, lime labels can be confusing so the following explains the various types:

Non-hydraulic lime: these are produced from limestones with no clay content and it therefore has no ability to set hydraulically.

Naturally hydraulic lime: these are limes with varying degrees of hydraulic quality, depending on the source. Blue Lias lime, produced in the West Country from limestones containing clays, is available in both putty form at NHL 2 and in dry hydrate form for other grades. St Astier, a French company, produces both non-hydraulic and naturally hydraulic limes. Building limes are becoming more readily available as the technology, knowledge and use of lime increases.

Dry hydrate lime (builder's lime): this is supplied bagged as a dry powder. It may be produced from both non-hydraulic lime and naturally hydraulic limes, but its use other than as a plasticiser in cement mortars is limited. Once hydrate of lime is produced and bagged it starts to reabsorb carbon dioxide and thus has limited ability to be used effectively in lime mortars as it has already started to revert to its limestone state and its use within effective mortars is therefore reduced.

There is no British Standard for lime-based mortars although there is currently a move to rectify this situation. Cement-based mortars do have several British Standards defining their use and components.

When used as mortar to form a wall, lime mortar acts sacrificially to the brickwork it bonds. All historic brickwork, as we have discussed, needs to breathe and lime mortar is vapour-permeable and thus moisture generated by weather (rain) evaporates more efficiently. In contrast, because cement mortars are impervious, water tends to concentrate within the bricks rather than within the overall wall (both bricks and mortar). This is particularly important to note in regard to historic brickwork. In softer fired, early bricks (fired at temperatures of about 850°C) the use of lime mortars ensures that such bricks, vulnerable to erosion, do not become adversely affected by heavy wetting from rain concentrated on the face of the bricks. There is a balance in lime-mortar walls between the mortar and the bricks, allowing wetting and drying to occur through evaporation. This is absent in cement-bound walls. The impervious cement mortar will not absorb rainwater run off and will concentrate additional water in the more absorbent brick. During winter this excess water will leave bricks vulnerable to frost attack, potentially resulting in erosion of the bricks due to the expansive action of ice crystal formation within their outer surface.

The use of lime mortar will therefore allow the whole wall to both accept and later evaporate water at a much more rapid rate, thus allowing the wall to dry out more effectively than where a cement mortar is used. A well-constructed brick wall using lime mortar will only require re-pointing after about 150–200 years and will not have to incorporate expansion joints due to the ability of lime mortar to be more flexible and self-healing or autogenous, than cement mortar. Autogenous simply means the ability of a material to fail under stress and heal itself automatically. Lime mortars are autogenous in that when stress-induced cracking occurs, perhaps as the result of seasonal movement, the crack exposes the inner mortar to carbon dioxide which induces more carbonation and thus self-healing of the crack. In contrast, a cement mortar will crack, allowing water penetration and ultimately leading to the potential for decay of the bricks or stones.

36.5   Damage to brickwork by use of cement mortars

The use of cement mortar in modern manufactured bricks is less problematic as bricks are now produced in frost-resistant grades so cement mortar usage is not likely to cause frost attack deterioration. Modern bricks are generally fired to higher temperatures in kilns where the temperature is better controlled throughout the kiln than in the early kilns or clamps previously used to fire bricks.

In summary, lime is generally appropriate for historic buildings, but must be used appropriately and appropriate investigative works must be undertaken prior to their use. The advantages are summarised by the Scottish Lime Centre Trust (1995: 1) as follows:

> The most effective methods of repairing traditional masonry buildings almost invariably involve the use of materials and techniques employed in their original construction. Traditional lime mortars are more permeable and more flexible than cement mortars, they contain fewer soluble salts, and they are environmentally more friendly than cement mortars. They also look better.

Sustainability of lime products

The sustainability of building products is an increasingly important issue with concern over climate change and environmental damage. Lime uses less energy in its production process than cement. For every tonne of cement produced its equivalent weight in carbon is also produced, making lime a more sustainable material.

Another important aspect of lime is that it holds components such as bricks or stones apart, as opposed to sticking them together in the way that cement mortars do. This means that components such as bricks or stones bound by lime mortar can be easily cleaned and re-used. However, where a cement mortar has been used the ability to re-use walling materials on later projects is precluded or reduced.

Other lime-based products

Lime was not limited to the production of mortars for brickwork and stonework. It was also used in the production of lime plasters and decorative lime washes as well as in many other industrial processes such as steel production and within the pottery production process. Industrial processes include its use in the production of paints and dyes, glass and soft metals such as copper and aluminium as well as for bleaching of paper and in the tanning industry. It has been used for generations in agriculture to stabilise soils. Other uses include the treatment of acidic effluents because of its powerful alkaline properties and also in the filtration of acidic flue gases.

Lime plasters

Lime plasters and renders have traditionally been used for the external and internal finishes of buildings and are used in the renovation of historic buildings. Lime plasters are mixed with coarse sand in the same way that mortars are, but the resulting surface finish is less smooth than gypsum plasters although the use of successive coats of lime wash renders them smoother with the passage of time. Gypsum plaster sets by hydration in a similar way to cement and has been used in the United Kingdom since the thirteenth century. It produces a much smoother finish than crude lime/sand plasters. However, this section will concentrate on the use of lime plasters.

Lime plaster may be applied to both masonry walls (on the hard) and to lathes, either on ceilings, walls or partitions. It is generally produced using non-hydraulic lime putty but can also be made using feebly or moderately hydraulic limes. It is generally mixed with coarse sharp sand to produce "coarse stuff" for first coats, applied directly to the masonry or onto lathes. Second or finishing coats are generally mixed using finer, softer sand to achieve a better, smoother finish.

As with all historic lime technology there is no exact science governing the methodology adopted by the tradesmen of the time as each adopted their own techniques dependent on the materials available and the specific requirements of the job in hand. So, although we may adopt an ideal or optimum technology in defining the craft of the plasterer it would, at the time of construction, have been a much more serendipitous result depending upon circumstances, skills and materials available. It is therefore essential when specifying repair and intervention work on historic plasters to carry out a detailed analysis of the original materials so that any new or repair work accurately matches the components of the original material. The same principle, of course, should be adopted when repairing historic mortars.

Historic lime plaster may have two or three coats dependent upon the quality required of the finished work. Vernacular buildings usually adopted a two-coat process whereas more prestigious, better-finished buildings would have had a three-coat methodology. For two-coat work there may have been a first or dubbing coat of lime and coarse sand in a generic 1:3 mix ratio and the top coat, setting, or finishing coat may have used a finer sand to achieve a smoother finish. In third-coat work the final coat is called "fine stuff" and is usually mixed with sharp, fine sand to achieve the necessary smooth finish. To assist a more rapid set, pozzolanic additives may have been incorporated in the same way as in brickwork mortars. (See Development of cement section, page 202.)

Lime plasters are not as cohesive as cement-bound renders and modern gypsum-bound plasters so animal hair is combined with the plaster at the mixing stage. This hair acts rather like reinforcing in modern concrete, improving the tensile strength of plasters. The type of hair used is usually ox, cow or goat, or sometimes horse hair. Human hair is not usually used as it is too fine and less strong than animal hair and is less resistant to the alkalinity of the lime mix. In terms of its use in the plastering process, base or dubbing layers may be reinforced with animal hair but subsequent layers may or may not incorporate hair reinforcing.

Where lime plaster is applied over lathes, such as for ceilings or lathed partitions, the lathes are usually riven hazel, oak or chestnut, although they may have been sawn. Reed may also have been used in replacement of lathes in areas where it is readily available such as in East Anglia. Riven lathes provide a rougher surface and plaster adheres better to them. Lathes should be placed about 8–10 mm apart to allow the plaster to squeeze through the gaps and form a mechanical key overlapping the backs of the lathes. Such "squeezings" through the lathes are referred to as "snots" and are a very necessary part of the way in which the plaster attaches to the lathes. In failed ceilings it is the break-up or detachment of these "snots" that causes plasterwork to become loose, eventually failing and dropping off the ceiling or partition. When plastering onto lathes it is essential to push the first coat well into the gaps between the lathes so that the "snots" extrude through, flop over and attach to the back of the lathes, thus holding the plaster to the lath by a physical "hook".

36.6   Demonstration panels on the top right demonstrate the "snots" or mechanical key

36.7   Lath and plaster wall showing "snots" and the lath construction

Stucco plasters and renders

Stucco was a very fashionable material during the Georgian and Regency periods of the eighteenth and early nineteenth centuries. Curl (1999) states that it is a slow-setting material and as an external render it comprises "lime, sand, brick-dust, stone-dust, or powdered burnt clay nodules, mixed with water, used as a finish instead of stone, often lined to resemble ashlar work and moulded to form architectural features such as string-courses, cornices etc.".

It was used by architects including Nash and the Adam brothers to imitate stone in areas where naturally occurring stone was not readily available. This style of building was constructed in the nineteenth century in London and elsewhere, particularly at seaside locations like Brighton where it had good salt-resisting qualities. Woodforde (1985: 23) states that stucco rendering was "generally restricted to the ground floor where it could simulate at trifling cost the rusticated masonry for ground floors of the Palladian houses". Thus, very simple and cheap buildings were given a grand status by the application of stucco (see Figure 23.1).

External stucco render has a number of variations, which are defined by Constantinides and Humphries (2003) as follows:

- **Common stucco**: an external render made using hydraulic lime, sand and reinforced with animal hair.
- **Rough stucco**: an internal plaster made from fine sand, very pure (non-hydraulic lime) and used internally to imitate stonework.
- **Bastard stucco**: made from non-hydraulic lime, fine sand, and finished by polishing to a very flat surface.
- **Trowelled stucco**: a non-hydraulic lime render, applied as a finishing coat, polished and painted with lime wash.

The use of stucco declined in the mid-to-late nineteenth century with the Gothic Revival of that period and the rise in popularity of brick buildings. It was also perceived as a dishonest material because its imitation of stone and such deceptive materials fell out of fashion in the later nineteenth century through the writings of John Ruskin and the work of architects including Pugin and William Morris.

Lime wash and harl

Lime washes as decorative and sacrificial coatings have been used for many thousands of years. They can also be used to stabilise loose surfaces and protect them from further deterioration. They are made from matured lime putty and applied in thin coats over a period of time, allowing each coat to dry out before applying the successive coat. It is a slow process and to try to achieve a heavy coat will be counterproductive to drying and over-coating.

Lime wash is a special material and has a characteristic 'glow'. This is because it has an ability to reflect light due to its dual reflective index, meaning light is reflected back at twice the source rate. Crystals of calcite (calcium carbonate) are produced when lime sets in air through the absorption of carbon dioxide, and it is these crystals that give lime washes this effect (Bennett, 2001).

The washes can be dyed with a variety of natural earth pigments to achieve a large range of colours. However, it is not easy to replicate mixed colours so it is essential that sufficient mixed colour is made in each batch to complete the decoration of a project or section of a project. Tallow, or raw linseed oil, is sometimes mixed with lime wash to improve water resistance, which is particularly applicable for lime washes applied externally.

In Scotland, lime harl is customarily used for the exterior of traditional buildings. This involves the inclusion of small aggregates in the mix, which is then applied to the wall giving a rougher texture. Such material successfully covered rubble buildings but the dressings around the window and door openings were sometimes designed to be seen.

An important aspect of lime washes is that they are essentially sacrificial coatings. This means that they allow weathering to attack the coating rather than the underlying substrate. It is therefore essential that lime washes are re-coated regularly to ensure that this sacrificial process and, thereby, the protection is maintained.

36.8   Lime harl, Bo'Ness

Conservation implications of use of lime

We have emphasised throughout this section the need to understand that historic buildings must be able to breathe in order to disperse any build-up of moisture within the structure. Lime as a binder is pervious and allows this breathing process to occur, and will not compromise the passage of moisture migration throughout a structure. However, the use of modern impervious paints and decorative coatings on historic buildings may compromise this breathability. It is also important to remember that only earth-based pigments should be used in production of lime washes. Modern dyes and colourants may adversely affect the permeable nature of lime and should be avoided.

However, although lime has considerable advantages for historic buildings it is not without its difficulties. The advantages and disadvantages of the material are summarised below.

Advantages:

- It facilitates the effective repair of old buildings whose original formation made use of lime technology to retain 'breathability'.
- It facilitates re-use of components.
- It permits long lengths of construction to be undertaken without the need for expansion/contraction joints – some ancient estate enclosing walls built in lime mortars are several miles long without expansion joints.
- It is flexible and to a certain degree self-healing if cracking does occur.
- Its lower strength makes it suitable for use with old materials such as soft bricks and stones.
- It is effective in resisting salt attack in masonry and brickwork.
- It has a better appearance when used to repair old buildings.
- It has improved resistance to rain penetration, allowing evaporation to occur more quickly than in constructions adopting cement binders.

Disadvantages:

- It is slow to harden and must be used within the 'growing season' of April to October, before frosts can adversely affect it.
- It is less strong in compression than cement mortars.
- It is more expensive than cement mortars and requires greater operative skills.
- Special equipment as well as specialised knowledge and skill are required compared with cement-based products.
- It must be protected and monitored during setting and requires a greater time commitment during use.
- It takes more effort to produce good-quality lime mortars and plasters.

36.9   Deterioration of lime harl, Dunblane

During the twentieth century, the use of lime reduced dramatically as cement-based products became cheap and readily available. Viewed as faster and easier to use, they quickly superseded lime and the impact of this change on historic buildings has been significant. The widespread use of harder cement products has caused deterioration of the softer stones it aims to bind and protect and water becomes trapped behind the non-breathable cement, causing erosion of the stone or brickwork. This is evident in Figure 36.10 where the stonework has seriously deteriorated due to the use of inappropriate, cement-based re-pointing.

An increasing awareness of this problem in recent years has resulted in the revival of the use of lime in conservation circles. Organisations like the Scottish Lime Centre Trust and in England the British Lime Association have investigated its use and introduced training for tradesmen and property professionals. Considerable research and training resources have been required to revive this lost knowledge.

Its re-introduction has not always been viewed as successful, however. In some cases, like the harling of Stirling Castle Great Hall in a bright coloured harl has caused controversy (Figure 36.11). However, this

36.10   Sandstone damage at New Lanark

36.11   Lime harl, Great Hall, Stirling Castle

perhaps highlights an education requirement of the public as well as property professionals and building operatives. People need to understand how historic buildings work and how they were originally conceived. For example, there is a fashion for bare stone buildings where the harling or render is removed and rubble walls which were originally designed to be hidden by politer harl or lime wash finishes are now exposed to the elements. The loss of this sacrificial coat is potentially detrimental to the stone and may encourage dampness. In exposed locations like the north and west of the British Isles, a lime coat offers protection from the elements and allows the building to breathe. Removal of this coat ultimately leaves a building much more exposed to the weather. In Figure 36.12, it is evident that the building on the right retains its original "harling" (the English equivalent might be pebble-dash) but the building on the left has had the protective coat removed. Such removal affects not only weatherproofing but also visual appearance, originality and authenticity. This scraping of applied original finish to expose underlying stone contributed to a reaction that resulted in the formation of the anti-scrape movement that eventually morphed into the Society for the Protection of Ancient Buildings (SPAB) in the latter half of the nineteenth century.

Despite the best efforts of the leading conservation bodies, lime has still not been as widely adopted as had been envisaged. Cement is still viewed by the building industry as a cheap and easy-to-use product and homeowners remain uninformed of the damage they are potentially doing to their historic buildings by using cement. Builders

36.12   Comparison of buildings with lime harl retained (right) and harl removed (left)

employed by homeowners, for the most part, are unaware of the damage that they may be perpetrating by the use of cement binders in historic buildings. While the use of modern cement products in new construction is accepted, the misplaced use of it in historic buildings is a major factor in the deterioration of historic fabric. Further education is therefore needed in this field.

Development of cement

Although lime concrete was known to the Ancient Egyptians and the Ancient Greeks, it was the Romans who developed its use to a fine art. The Romans had created constructional forms that did not rely on specialist skills in order that buildings could be constructed rapidly and by a less skilled workforce. Although lime was used, the Romans also developed an early form of concrete that permitted construction of large and complex structures in Rome, including the Colosseum and the Pantheon. Examples outside Italy include the Pont du Gard in France, and at many fortifications in Britain such as the walls at Colchester.

The Romans discovered concrete serendipitously. While undertaking construction at a place called Pozzuoli, near Vesuvius in Italy, they used a pink sand-like material which caused their mortars to set very rapidly and with greater compressive strength than usual mortars formed using ordinary sand and lime. The apparent sand was of course volcanic ash generated by the nearby volcano. This ash had certain elements within it that caused the mortar produced with it to set hydraulically. This created a very strong mortar that set rapidly and with improved qualities of compressive strength. Hence the term pozzolanic additive, is used in lime technology to describe any material that, if used in mortar production, produces an increase in rapidity of set and creates stronger mortar. Vicat (1997) states: "Pozzolanas are essentially composed of silica and alumina, united with a small quantity of lime, potash, soda and magnesium."

Examples of pozzolanic additives include crushed brick or tile dust. These crushed products are derived from clay containing silcates and aluminates which are heated as part of the firing process to produce the original bricks and tiles. The process of heating is probably synonymous with the heating process adopted by Aspdin in 1824 when he discovered such qualities in heat-treated clay in his production of Portland cement.

Aspdin found that by combining finely ground limestone, or chalk, with clay and heating it to a certain temperature, a material that set with water was produced. It was the silicates and aluminates within the clay that allowed the cement to set hydraulically in the presence of water.

The particular setting properties of concrete made it an attractive material, particularly for use by less-skilled operatives during the early-to-mid-twentieth century when there was a need to rapidly produce new housing, especially immediately following the Second World War. It had also proven itself in the construction of water-retaining structures such as canals and docks during the mid-nineteenth century. The use of cement also removed the need for building processes to be undertaken seasonally as was the case with lime. Lime could only sensibly be used during the growing season of April to October (when the air is drier), and depending on the level of hydraulicity of the lime used. However, with cement, construction could be undertaken all year round.

Although cement has facilitated some twentieth- and twenty-first-century structures that could not have been constructed using lime technology, cement and concrete are not without their problems, including the deterioration of modern concrete buildings. The technology of cement and its composite material concrete was developed over the early and mid-twentieth century, and many buildings have suffered from inadequately understood use of the material, particularly where concrete is used with steel reinforcing to improve the tensile strength of structures. This is perhaps most prevalent during the middle years of the last century and more so in the immediate post-war period when construction had to be completed quickly to restore or replace buildings destroyed by Blitz bombing and in order to satisfy a rapidly expanding economy.

During the 1950s and 1960s there was inappropriate use of additives to improve set times and also a reduction in the cement content to save money. This has resulted in many of the buildings of this period suffering from premature decay. Probably the worst examples of premature decay is associated with the use of high-alumina cements. These products facilitated rapid setting times and therefore early re-use of expensive shuttering. However, an increase in the porosity of the concrete made using high-alumina cement, coupled with inadequate cover of the reinforcement by concrete, has resulted in the corrosion of reinforcing steel, threatening the longevity of some structures.

A further problem is that the aluminates contained in cement renders concrete subject to attack by soluble sulphate salts which may be present in soils or in hardcore laid contiguous with the concrete. In the presence of water, such as ground moisture, there can be a reaction with the tricalcium aluminate within the concrete to form etringite. Formation of these crystals within concrete causes expansion and cracking, thus letting in more moisture and exacerbating the sulphate attack. In early concrete structures this process was not fully

36.13   Concrete disrepair

understood, so numerous concrete structures suffer from this type of damage. Similarly, alkali-silica-reaction (ASR) creates damaging conditions within concrete. ASR is caused by alkalis present in cement reacting with certain types of silica aggregates to form a gel-like substance that expands and causes crack damage to the concrete containing it. This type of damage to concrete buildings can be extensive and may affect the entire structure, requiring major, extensive and invasive intervention.

While the focus of conservation bodies has been on traditionally constructed buildings and those pre-dating 1900, there is increasing awareness of the importance of more recent buildings and particularly those which were at the leading edge of design and use of new materials. However, if such structures are of historic importance then the basic principles of minimum intervention and minimum loss of fabric may be at odds with the extent of work necessary to return the buildings to a stable condition. New techniques have, fortunately, been developed as conservation philosophy has advanced to take account of the implications of advances in built-environment technology. However, the problems of inadequate knowledge of early concrete technology are potentially threatening our contemporary built heritage and this may become an increasing problem in the future if further research is not carried out.

Chapter 37

Roofing materials

In recent decades concrete roof tiles, profiled steel sheeting and felt have tended to replace the more traditional thatch, slate, stone and lead. These replacements are mass manufactured and generally cheaper than traditional materials. However, with over 400,000 listed buildings in the United Kingdom, the majority of which were built pre-1900, it is vital to understand the roofing materials used on traditional buildings. We will therefore examine thatch, slate/stone and metals in turn.

Thatch

Thatch or thack is a locally derived material which once covered all but the most prestigious buildings in the United Kingdom. The term may include straw, reed, heather and turf and sometimes a combination of these materials. Cheap and readily available, it was easy to repair and replace and so was the favoured material on domestic buildings for many centuries. It also did not require particularly specialised equipment in order to lay it. However, it is likely that it was no longer used for the most important buildings from around 1200 (Clifton-Taylor, 1987). It was also gradually discouraged in towns and cities because of its highly combustible nature, which, when combined with timber-framed buildings, meant that fire was a continual risk. This was particularly relevant in urban situations where the high density of buildings permitted the rapid spread of fire. This risk was exacerbated by the presence of certain trades such as blacksmiths, bakers or any operation involving heat and fire. The time from the medieval period up to the seventeenth century witnessed many severe town fires, the most well known being the Great Fire of London which started in Pudding Lane on 2 September 1666. Local byelaws and acts therefore sought to encourage the use of less combustible materials like slate and tiles for roofs and brick or stone for walls.

In terms of geographic distribution, thatch can in theory be found anywhere in the United Kingdom, but its use continued the longest on humble vernacular buildings. However, many of these have now been replaced with slate, although older thatch may still exist hidden under more recent corrugated iron coverings. The continued use of thatch is probably most strongly associated with East Anglia and the West Country, places where earth buildings still survive and locally available reed was easy to source.

The most common types of thatch are:

- heather
- straw
- reed.

There may be other materials present such as clay and turf which help to make the thatch waterproof. The thatch may also be weighed down with stones and ropes and/or protected by netting.

The different thatching materials used vary in their longevity. Reed thatch is the most durable and, although more costly to lay, may last for 60–80 years, whereas a straw thatch roof may only last 20–30 years. Thatchers developed their own particular styles and so local features emerged as a result. For example, a characteristic or signature way of decorating the ridges. Some of these can be quite elaborate in the designs used, the most favoured decoration being half-circles and points arranged in identifiable ways (Figure 37.1).

37.1   Local thatching style

Utility buildings such as barns tended to have straight ridge designs as did some churches in order to avoid excessively exuberant designs that might be associated with dwellings.

Buildings with thatch roofs generally do not require guttering systems but instead have very steep roofs of around 50°, helping rain to run off and thus avoiding premature rotting of the reed and discouraging moss growth. Moss is acidic and can seriously reduce the life of reed and straw thatch by encouraging saturation of the thatch. The ridge is the area most vulnerable to water penetration – turf, sedge or clay may be used in order to improve the water-shedding abilities of the roof. This is evident in Figure 37.2 where the ridge has disintegrated, allowing water to enter and rot the thatch at the vulnerable ridge and wind to lift and destroy the thatch.

Aesthetically, thatch of all types remains a very appealing building material, and in terms of performance it makes the building cool in summer and warm in winter. It has the added benefit of having good sound insulation properties. Thatch has particularly good thermal performance and does not require additional insulation material to achieve contemporary thermal insulation standards. It is, however, vulnerable to fire, particularly at junctions of thatch with chimneys. Modern building regulations may require that the thatch be underdrawn with a fireproof board.

There are also other problems associated with thatch, including damage by nesting birds and rodents. Walker et al. (1996: 8) suggest that problems from birds and vermin can be caused by incompletely threshed straw which has been an increasing problem since the use of combine harvesters.

37.2   An agricultural building where the sedge ridge has disintegrated

37.3   The same building as Figure 37.2 with the roof re-thatched – note the fairly simple ridge patterning

Thatch: conservation implications

Although thatchers may be available in certain parts of the country, such as Norfolk, other areas may have seen a decline of skills, such as in Scotland where thatched buildings are now less numerous. The loss of traditional craft skills has implications for the conservation of traditional techniques and may result in the loss of regional characteristics. Walker et al. (1996: 11) state that in any training courses there must be a focus on regional aspects or this could "accelerate the decline [of regional characteristics] rather than arrest it". The importing of techniques from one region to another is therefore a significant concern for traditional thatching.

Other issues relate to the lack of availability of suitable materials. For example, it is becoming increasingly difficult to source good, well-grown natural reed from wetland areas such as East Anglia or other

37.4   Extensive fire damage caused by a thatch fire in a seventeenth-century house in Norfolk

areas such as on the Tay estuary in Scotland. Many re-thatching projects use reed from East European countries and such reed may be inferior to good United Kingdom-sourced material. Reed sourced outside the United Kingdom increases transportation costs and the associated carbon footprint and is not, in conservation terms, a sensible alternative. Reed-bed management at sites within the United Kingdom improves sustainable sources and encourages investment in local production.

However, pollution of reed beds with nitrogen is a problem, as is the case with straw, and its impact is to weaken the material and make it unsuitable for thatching (Walker et al., 1996). Changes in farming practices have therefore had a significant impact on the availability of good material and the use of imported, less hard-wearing reed has resulted in reduced longevity of reed thatch.

Other issues such as difficulties in obtaining insurance for properties with thatch roofs and the higher cost of maintenance make it perhaps a less appealing prospect than its wonderfully aesthetic appearance should. Despite this, it is a very enduring feature and where a property has been traditionally thatched this should be maintained.

Slate and stone roofing

Slate is a metamorphic rock which had its origins as sedimentary rock and is therefore lamellated in structure, being made up of many thin, parallel layers (Centre for Conservation and Urban Studies (CCUS), 2000). In terms of performance, slate is most important for roofing, although it is used for walling and cladding, steps and stairs, paving floors and also for damp-proof courses. Although brittle, it is often extremely hard and close in texture, non-porous, quick-drying and frost resistant. It is also able to withstand atmospheric pollution.

Slate is found notably in Wales, the Lake District and the Highlands of Scotland, as well as the Isle of Man and the English granite areas of Cornwall, Devon and Leicestershire. The wide variation in the geology of these districts means that the slate varies considerably from location to location. It is important to understand these variations because they have a major impact on the resulting design of roofs. The following summaries give a brief account of the differences across the country:

Lake District: here the slate is more like a sandstone or limestone, being of volcanic ash. It is rough, very heavy and varies in colour, although is often greenish in hue.

Cornwall and Devon: the majority of Cornish slate comes from the quarry at Delabole. The slate is fine-grained, compact and strong but not heavy. It is grey tinged with green in colour. Although Devon produced many slates in medieval times, it is less hard and less enduring than Cornish slate.

Leicestershire: pre-Cambrian slates were used for many centuries. Even the humblest cottage had them in the Georgian period, although these local slates were superseded during the nineteenth century by Welsh slate which could meet the demand and therefore captured the market.

Wales: Welsh slate is strong, reasonably durable, completely non-porous and available in standard sizes. Its main advantage is that it can be split into very fine laminae, so it is thinner, smoother and weighs less than other slates. The supporting walls can therefore be less-strongly built and wooden rafters and battens lighter. The thin slate can also lie on low-pitched roofs of 22–26°. The machined look of Welsh slate is because of its geologic structure, which means it can be trimmed into standard rectangles. Also, since the end of the eighteenth century it has never been practice to lay Welsh slates in diminishing courses as with other natural slate. Instead it is laid in "tally" where the same size of slates are used for the whole roof, which is a simpler method of laying (Figure 37.5).

Use of Welsh slates, as a result of improvements in transportation such as the canal system, became widespread during the reign of George III. Huge quarries at Penrhyn and Blaneau Ffestiniog supplied the whole of the United Kingdom with slate material, resulting in a uniform grey roofing material throughout the country.

37.5 Welsh slate

Scotland: Scottish slate is a rugged and distinctive material yet has local variations depending on the quarry it was sourced from. The main quarries were at Easdale, Ballachullish, Aberfoyle and MacDuff, with the largest production being on the western coast quarries of Easdale and Ballachullish. Smaller local quarries were located along the Highland fault line through Perthshire. Scottish slates are laid random in diminishing courses. That is, the largest slates are laid at the eaves level and they gradually diminish in size with the smallest slates being laid at the ridge. This maximises the use of all the material in the quarry as even small slates can be used and are not wasted. The thick and rugged slates are laid on a steeper roof pitch and are laid onto timber sarking boards rather than battens. The slater requires greater skill to lay random slates in diminishing courses because he has to carefully choose the slate sizes.

Scottish slate has a particularly distinctive appearance which lends character to the towns and villages where it is used as the main roofing material. CCUS (2000: 6) state that its main characteristics are that it has variety and ruggedness, and it gives a visually enhanced pitch resulting in a "strong visual presence in the overall architectural compositions which make up the historic townscapes of Scotland" (Figure 37.6).

37.6   Scottish slate

Slate: conservation implications

The United Kingdom slate industry has declined significantly since its height at the end of the nineteenth century and early twentieth century. For example, since the last Scottish slate quarry closed in the late 1950s, it has only been possible to obtain such slate from second-hand sources. This has implications for historic buildings as effectively, any slate of Scottish derivation needed for new work, repair or renovation must have come from the destruction of another building. A further issue is that because Scottish slate is less widely used now, the skills to lay such a roof are reducing and the expertise is gradually being lost. Indeed, this loss of skill is not confined to slating. The National Heritage Training Group's *Traditional Building Crafts Skills Report: 2007* states that a further 5,000 skilled tradespersons would be needed in Scotland to provide for the necessary level of tradespersons, particularly those skilled in stonemasonry, carpentry and roofing. Although there have been research studies into the re-opening of Scottish slate quarries, to date this has not happened.

To a large extent the requirement for slate for roofing purposes is now being met by the introduction of foreign or so-called "Spanish slate", slates which are brought to the United Kingdom from Europe and China. These are cheap and readily available but in terms of appearance they are often considerably poorer than slates sourced from the

37.7   A cottage with a roof recently renewed with foreign or "Spanish" slate

United Kingdom. They are generally thinner and smoother with an extremely uniform, even shiny, appearance. Their durability is uncertain and it may well be that their longevity is nothing like that of home-sourced slates. Certainly in appearance terms they are very much the poorer cousin, but for building owners their lower cost makes them extremely attractive. However, the implications for conservation and the future integrity of historic buildings which have traditional slate roof coverings are significant (Figure 37.7).

Stone roofing

Stone slates are derived from stone that can be cut thinly. Morriss (2001: 104) suggests that the correct terminology for this type of roof covering should be "tilestones".

Stone roofing particularly refers to limestone sourced from the Cotswolds and Northamptonshire, although tiles derived from sandstone are also found. For example, in Caithness and Orkney large sandstone slabs are used on the roofs. Like Scottish slate, these stone slates are also laid in diminishing courses, but the pitch of the roof will vary depending on the nature of the stone used. Attractive and rugged in appearance, these roof coverings sit well on the stone buildings of the Cotswolds and other areas where they are used.

The use of such a thick and dense material means that these stone roofs are particularly heavy compared to other roofing materials and

37.8   Caithness slate

37.9   Stone roofing, Duddington, Northamptonshire

37.10   Graduated stone roofing, Duddington, Northamptonshire

therefore clearly will require a stronger timber roof structure to support them. Where the slabs are large the roof will also be of a low pitch.

Metal roof coverings

Historically, a variety of metals have been used in building construction and some have proved to be particularly useful for roof coverings, including lead, copper, zinc and corrugated iron.

Lead is a malleable material which has been used extensively for roofing and for waterproofing of joints, including flashings and valley gutters, for centuries. It had additional uses for roofing for rainwater goods and their fixings. Historically, it was used from medieval times for windows as glazing channels or "cames", and it also proved to be useful for plumbing. However, as a roofing covering, if laid well, it is an effective material and can be used for low-pitched or flat roofs, usually in the form of large sheets. It is long lasting, particularly compared to modern alternatives.

According to the Lead Sheet Association, if correctly fitted and specified lead roofing can potentially last for 100 years. The Association states that "the chemical composition of rolled lead sheet is strictly governed by the provisions of BSEN 12588:1999, which effectively control the grain structure to make the lead sheet more resistant to thermal fatigue without affecting malleability" (Lead Sheet Association, 2008).

Copper and zinc are also used for roofing. Copper can be a pleasing roofing material, producing a green finish due to patination. However, they are less commonly used in the United Kingdom than lead and may prove expensive as a roof covering.

In certain parts of the country, such as in rural Highland Scotland, corrugated iron is used extensively as a cheap and readily available roof covering. It has been used since the 1820s but became popular from the 1850s. The earliest-known corrugated iron building in Scotland is the 1851 ballroom on Balmoral Estate (Thomson and Banfill, 2005). It is therefore very much a material of the nineteenth-century Industrial Revolution, with the mass production of materials like this occurring for the first time.

Although widely used for farm buildings, corrugated iron is also used to cover domestic buildings, possibly where they have traditionally had a thatch roof. Corrugated iron was also used to construct entire buildings. It was cheap and kits were readily assembled – even two-storey houses or churches could be purchased. Thomson and Banfill (2005) state that corrugated iron makes a number of important contributions to the historic environment:

37.11 A house with a lead roof

- aesthetic qualities in terms of its colour and texture
- indicative of social change
- reflects industrial history of pre-fabricated materials
- represents design innovation.

37.12 Corrugated iron roof on a domestic building which was probably originally thatched

37.13   Corrugated iron church, Killin, Perthshire

Metal roofs: conservation implications

Metal roof coverings – particularly lead – are still used, although their use is increasingly being replaced by modern sheet metals and felt roofing, largely on the basis of cost and the lower skill level required for laying. However, the longevity of lead and other traditional roofing materials needs to be appreciated as well as the importance of maintaining authenticity for historic buildings and potentially the sustainability implications of using materials which have a significantly longer life span.

More recently, the potential significance of corrugated iron has been gaining recognition for its important contribution to vernacular architecture, yet few corrugated iron buildings are listed and many remain undervalued for their aesthetic and design qualities. With limited protection they remain vulnerable to damage or demolition. Further research and investigation is needed in this area.

Clay roof tiles

Although slate gradually replaced thatch, in certain areas tiles became the favoured material. These are usually made in brick kilns, but the difference between them and brick manufacture is that the clay is mixed more carefully and baked harder to give them a more uniform colour and texture. In England they have been made in the south and east counties since the thirteenth century, and baked floor tiles were used extensively in the thirteenth and fourteenth centuries. However, until the development of the canal system in the nineteenth century their geographical distribution was limited to the south and east. The opportunities offered by this extensive transport network meant that Staffordshire and Shropshire became two of the most important areas for tile production.

The main types of tiles are outlined here:

Plain tiles (both nibbed and un-nibbed): these do not overlap with adjacent tiles so must be laid in "double lap" to give two thicknesses of tile. The underside usually has torching, a mortar pointing, underneath. Although the pitch of the roof may be 40°, a steeper angle of 50–60° can be more aesthetically pleasing and in certain locations, such as Kent, "cat slide" roofs which almost reach the ground are commonly found. An accepted minimum pitch of 35° ensures water tightness and prevents water penetration due to capillary action through the tile joints as well as reducing wind-lift damage. Nibbed tiles are hung onto battens with a nib that is part of the tile itself. They may also be nailed or pegged. The use of nails or wooden pegs is particularly relevant in un-nibbed plain/peg-fixed tiles as in the south-east of England.

Pantiles: these were originally imported from the Low Countries, probably as ballast when ships were returning otherwise without cargo, and so are particularly prevalent in the east of England and Scotland, the main trading locations with that part of Europe. However, they are also found in Somerset where Bridgwater was a main producer.

Pantiles are larger than plain tiles and are normally S-shaped in profile. This means that they are not as close-fitting, so mortar torching underneath is required to ensure water-tightness. Their complex shape means that they are less suitable for roofs with valleys or hips and an alternative material such as lead or zinc must be used in valley construction. However, being laid single-lap, the roofs are lighter so less timber is required in the roof structure. They may also be laid at a pitch of 30–35°.

Early tiles exhibit local variations as locally sourced clays would have resulted in different colours of tiles. Styles of tiles used also vary locally and depending on the local architecture and manufacture. In Scotland they are common in Fife and the Lothians where the lower layers of the roof are sometimes in slate which may be used to allow

37.14   Plain tiles

the rainwater to flow more evenly into the guttering system (Figure 37.15). However, like the mass production of the brick-making industry in the nineteenth century, gradually the local influence of tile manufacture was replaced with a more uniform, machine-produced approach.

Clay tiles: conservation implications

The distinctive landscape of pantiles and clay tiles in many British towns and the countryside offers a variation in hues, textures and roofscapes which make a significant contribution to both urban and rural architecture. However, the replacement of these tiles with modern concrete substitutes can significantly reduce the value of these buildings. It can result in a significant loss in the aesthetic value of a building, as well as its historic integrity. While inevitably some replacement of traditional tiles may be necessary through damage or deterioration, any work undertaken requires careful consideration.

37.15   Pantiles with a slate base, Fife

Chapter **38**

Walling and structural materials

Within the United Kingdom there are many different types of walling materials. Like other building materials, these vary considerably across the country, influenced by local geography and geology. The main materials are earth, timber, stone and brick. These are examined in turn.

Earth construction

Vernacular buildings are derived from their local area and no building type more typifies this than earth buildings. Houses were constructed using locally dug materials and basic skills until the nineteenth century. Clifton-Taylor (1987: 287) describes these "unbaked earths" and states that they are "generally the stuff of rather humble build-ings, mostly cottages, small farm-houses and their appendages". His-torically the use of earth has been of considerable importance, the largest earth structure in the United Kingdom being the Antonine Wall built across central Scotland by the Romans. Although earth is no longer a dominant building material, it remains important in other parts of the world. Nother (in Keefe, 2005) states that at least 30 per cent of the world's population lives in buildings of unfired earth, particularly in hot climates and where timber is lacking.

Despite their prevalence in warmer climates, many survive in the United Kingdom but tend to be geographically concentrated in particular areas, strongholds being East Anglia and the southwest of England. However, they are also found in Scotland, particularly Morayshire and in the Carse of Gowrie in Perthshire (Figure 38.2), as well as in the English counties of Buckinghamshire, Oxfordshire, Lincolnshire and in the East Midlands. Within these regions there are variations depending on the particular favoured method of construction in each area and reflecting available local materials and skills.

There are several advantages to this method of construction, notably that these buildings have excellent thermal and sound insulation properties, making them cool in summer and warm in winter. Construction costs are minimised by using a raw material which is both plentiful and potentially free. Earth construction also uses less-skilled labour than that required for say, a masonry wall, and tools do not have to be particularly specialised.

However, they are not without their problems, particularly that of a wet environment which may erode the walls, especially if not properly maintained. Also, animals may burrow into the walls, causing damage.

The typical features associated with earth buildings include small windows and hipped thatched roofs with a large overhang to throw the water away from the walls. Thatch is a lighter material than other roofing types and is therefore eminently suitable for earth buildings. Thatch may be straw or reed depending on what is available locally. However, the "golden rule" for earth buildings as outlined by Clifton-Taylor (1987) is that they must be kept dry or they will disintegrate. However, if the walls are kept free from damp through a good overhanging roof and a stone or brick plinth they will remain strong and durable for centuries. The old adage, "A stout pair of boots and a good hat" exemplifies this.

There are several methods of clay or earth construction, including use of turf, but the main ones are outlined below:

Clay lump or "adobe": examples in the United Kingdom generally date from the late eighteenth or early nineteenth centuries and the method is thought to have been introduced from abroad. Clay lump buildings are found predominantly in East Anglia – surviving buildings may be 300 years old. The method of construction involves mixing clay with straw then pressing it into wooden moulds which are dried in the sun. These blocks are then built-up in a method similar to a block wall but the mortar is of puddled clay. The finished walls are usually rendered with protective coatings like tar and "ruddle" or red ochre which may then be lime washed. This type of construction may also be hidden behind a façade of brick or flint.

Figure 38.1 shows an example of a rammed earth or clay lump, agricultural building in Norfolk, now disused, with tarred coating to protect it. The brick and flint plinth was adopted to provide protection from ground water. This building is within a known flood plain and the height of the plinth might have been adopted to protect against historically known flood levels.

38.1   Rammed earth or clay lump, an agricultural building in Norfolk

Cob (or Pug in chalk areas): in this method mud and straw are mixed in horizontal layers, allowed to dry out and then pared down. In this way the wall is built-up in layers, the straw adding strength and helping to reduce cracking and shrinkage. Structures built this way are found in Devon and southwest England, as well as in north Wales. In the Oxford/Aylesbury area where this method uses chalk and the end result is yellow in colour this method is known as "witchert". Cob is a more labour-intensive construction process and, for example, it may take up to two years to construct a two-storey house (Clifton-Taylor, 1987).

In Scotland a mudwall system was adopted using straw and clay built on a stone plinth and the constructed walls were then pared down. Large stone boulders could also be incorporated into the wall. This type of construction can be seen in the Inverness area, Aberdeenshire and notably in Perthshire where the mid-eighteenth-century Old Schoolhouse at Cottown, Errol has been recently renovated by the National Trust for Scotland (Figure 38.2).

Pise de terre: this is a rammed earth method using a shuttering system – the buildings tend to be much stronger as a result. This method, together with clay lump, was introduced to Britain in the 1790s as part of general agricultural improvements, but despite this pise de terre was never a popular method of construction (Nother, in Keefe, 2005).

38.2   Old Schoolhouse, Cottown, Errol, Perthshire

Earth: conservation implications

Keefe (2005: 31) states that "the durability of an earth wall will be dependent as such upon the raw material originally employed as it is on the degree of care with which the building was constructed".

While apparently simple, it is vital to understand the construction of particular earth buildings, especially if any repair works or interventions are to be undertaken. It is particularly important to avoid the use of modern materials like cement renders which can cause serious deterioration in an earth wall. These can trap moisture, prevent the building from 'breathing' and can ultimately cause the wall to fail. For example, the Old Schoolhouse at Cottown, Errol in Perthshire (Figure 38.2) was in serious disrepair in the early 1990s due to poor and inappropriate repairs. Reen (1999) lists the various problems as including water penetration to the earth wallheads due to poorly maintained thatch, internal concrete floors, poor site drainage and impervious cement renders to the external walls together with modern gypsum plaster on the interior. This all resulted in moisture being trapped within the earth walls which in turn encouraged infestation by rats and birds and growth of vegetation. This example indicates how an earth wall can be vulnerable if ill-considered modern interventions are undertaken. The building was, however, successfully renovated by the National Trust for Scotland and lime plaster and lime washes were used to allow the building to breathe.

It is evident that, like all buildings, good maintenance is essential, including maintenance of site drainage, removal of vegetation and ensuring that rodents and insects such as bees are not burrowing into the walls. A stone plinth may also be required to prevent rising damp but modern injected damp-proof courses should be avoided as again, these alter the drying-out process in a wall and can result in more problems than they try to solve. Earth buildings have a potentially long life if properly cared for.

Wattle and daub

The construction of post and wattle buildings where timber posts were interwoven with branches, twigs and reeds have been used since Romano-British times (Wright, 1991). However, it is more commonly associated with timber-framed buildings. In this case the mud and straw are pressed into interwoven wattle or withey panels, usually constructed of 'green' hazel which is a lightweight wood but which is also tough and pliable.

The daub is a mixture of clay, straw and dung. The origin of the incorporation of dung in the mix is open to interpretation. It probably resulted from using animals to 'tread' the mixture to achieve the right consistency in large quantity and the resultant dung or droppings (including urine, a known antiseptic) became incorporated within the daub mix. The constituent parts of animal dung contain lignin (vegetable fibre) which, probably serendipitously, resulted in a more cohesive mix.

The building would then be coated with lime wash which allowed the building to breathe and the damp to evaporate.

Wattle and daub: conservation implications

In terms of conservation, wattle and daub panels can be easily repaired if they have become damaged. This can be clearly seen in Figure 36.6 showing a demonstration panel of wattle and daub. Note the vertical "staves" through which the witheys or wattles are horizontally woven – rather like the warp and weft of cloth weaving. The daub mix was, probably, hurled or thrown at the wattles to ensure good penetration and mechanical attachment to the witheys. (Note: the top right-hand panels in Figure 36.6 are of lathe and plaster and are therefore not indicative of association with wattle and daub.)

Daub is susceptible to damp and mechanical damage. If the daub is lost then both rot and insect damage to the wattles is a hazard. Painting over with modern impervious paints must be avoided in order to allow the panel to breathe. External render repairs should be carried out using a similar method to the original construction and materials used. Any applied render must use lime-based renders and breathable paints such as lime wash.

Timber

The wide availability of timber means that it was an obvious choice when constructing shelter. Simply constructed wooden shelters offer indigenous peoples with cheap and effective shelter (Figures 38.3 and 38.4). However, as a highly versatile material, it also lends itself to very sophisticated construction.

Timber-constructed vernacular and civic buildings were prevalent from pre-history until its use declined in the sixteenth and seventeenth centuries. Nonetheless, it has seen a revival in modern thermally efficient forms of construction, but is limited in its ability to be used in multi-storey structures above five storeys. However, its susceptibility to damage by fire, rot and insect infestation means that only limited medieval timber buildings survive.

Timber can be divided into two main types of hardwoods and softwoods.

38.3   Simple wooden huts (in teak) providing living and working accommodation for a hill tribe in northern Thailand

38.4   A group of village huts for the Karren people of northern Thailand

Hardwood

Certain hardwoods are native to the British Isles such as oak, beech and elm. Dense and slow-growing, these timbers are more resistant to attack from weather and insects. Although in abundance until the medieval period, their availability declined thereafter.

The favoured building material was oak which is found growing throughout Europe, America and Asia. It is dense and straight-grained, the hardest and most durable species being English oak. It was used for centuries for construction of dwellings, roofs and for ships. It does not need to be painted and weathers to a silvery grey colour with time. Although oak is a remarkably enduring form, it does have problems. If used in its green form (as it invariably was) it moves significantly. Oak – as opposed to other forms of hardwood – seasons very slowly, so a 300 mm section of oak timber will still be seasoning for about 30 years at the very least. It seasons at a reducing rate dependent upon the thickness of section. For example, a 300 mm square section of timber, if cut open after 100 years will start to season afresh after cutting. So movement due to shrinkage is inevitable and explains why so many of our oak-framed buildings can appear so distorted.

In addition to native species, during the nineteenth and early twentieth century, exotic hardwoods were imported, many from the extensive British Empire. Mahogany, teak and walnut were used generally for high-quality interior work. However, with increasing concern in

38.5 Buildings in Tombland, Norwich (showing distortion of timbers)

recent years about deforestation of irreplaceable tropical rainforests and environmental sustainability of such ecosystems, the popularity of such exotic hardwoods has waned.

Softwood

These faster-growing trees are native to northern and central Europe – including Scotland – and include the fir and pine species. These are less resistant to decay or attack and so require a protective coating, usually of paint. Baltic pine and Canadian red pine were commonly used but the most favoured material for construction was pitch pine which is a native of the southern states of America such as California and Florida. It is a dense wood but is highly resinous and so difficult to work.

Although oak was the dominant timber used in England, some quantities of pine were imported during the medieval period. Clifton-Taylor (1987: 296) states that during the thirteenth century fir was imported from Norway for royal works although this material tended not to be used for construction purposes. With the decline in the use of oak from the late sixteenth to early seventeenth centuries, it was replaced by imported softwood timber from the Baltic and Scandinavia during the seventeenth and eighteenth centuries. During the nineteenth century Canada and America also became significant exporters of softwoods.

Today, softwoods dominate the building industry. Fast-grown on plantations, these are very different timbers from the pitch pine and similar species of earlier centuries. In terms of quality the speed of their growth means that they are susceptible to rot and damage from insect attack, requiring additional use of protective chemicals and paint products.

Damage to timber

As an organic and natural material, timber is vulnerable to attack from a number of sources. There may be fungal attack through wet or dry rot and insects such as wood-boring beetles can cause considerable damage. Certain conditions will encourage the growth or infestation, notably damp. Timber needs to be carefully maintained if it is to have a long and useful life as a building material.

Fungal attack is of two main types, wet rots and dry rots. Wet rot is commonly found on the surface of timbers and may be noticed in areas prone to wet conditions such as window-sills and the base of door frames where they are in contact with the ground. It is caused by a fungus (*Coniophora puteana* for example) and occurs where the moisture content of the timber is high. Where wet rot occurs it is usually fairly well confined and can be treated by the removal of the section and replacement with a suitably fitted new piece of wood.

Dry rot is potentially more damaging. It is also caused by a fungus (*Serpula lacrymans*) and occurs where there is little air movement and the timber moisture content is above 20 per cent. It may therefore occur below floorboards or in roof spaces where there is limited movement of air and is often caused by poor maintenance such as leaking valleys or gutters. Where timber joists are sitting on or close to the damp the timbers can become infested.

In the right conditions, the fungus develops a fruiting body which spreads out the spores, often orange in colour, and white hyphae or threads spread through stonework, plaster and brickwork, sourcing food and moisture from the available timbers. Where there is a considerable mass of hyphae these are seen as a white cotton-wool-like texture and are referred to as mycelium. The affected timber will become dry and heavily fissured and if badly affected will become so brittle that it can crumble and disintegrate and totally lose any structural integrity. This obviously has serious implications for structural members such as joists and roof timbers as the structural stability of the building may be compromised if not treated.

38.6   Wet rot fungus

38.7   Damage to external timber by wet rot

Historically, whole-scale destruction of interiors was undertaken where dry rot was discovered. It was believed that extensive areas of sound timbers needed to be removed in order to ensure the integrity of the building. This often resulted in the burning of affected timbers and extensive use of chemicals injected into the walls to kill the fungus. However, it is now known that such destructive works are generally not required. The fungus requires certain conditions to grow and if the source of the water ingress is repaired, such as a leaking gutter, and ventilation is introduced the fungus will not thrive. While damaged timbers are likely to need replacement or repair, the removal of large sections is unlikely to be necessary unless the fungus is extremely extensive. Some localised treatment may be necessary and specialist advice is generally required to find the best course of action.

In addition to fungal attack timber is also susceptible to attack from various insects. The type of attack will depend on geographical location and the type of timber used. Certain timbers, particularly softwoods, are more susceptible than others, and conditions such as being warm and damp can also encourage infestation.

The most common infestation is by the common furniture beetle (*Anobium punctatum*), the larvae of which bore through the timber leaving a honeycomb of small holes. They require treatment by application of a suitable chemical, although total eradication where infestation is extensive can prove difficult. However, where the infestation is localised the damage to the timber is unlikely to compromise the

38.8   Timber joist damaged by wood-boring insects

timber's structural integrity. More problematic is the death-watch beetle (*Xestobium rufovillosum*) which is found in the south of England. It tends to favour larger timbers but is extremely destructive and unlike the common furniture beetle, infestation may affect the structural integrity of timbers in a building. Again, seeking specialist advice from a consultant experienced in dealing with timber decay in historic buildings is advisable.

Timber: conservation implications

During the twentieth century the availability of hardwoods declined and commercial plantations focused on fast-grown softwoods. The result is that modern buildings utilise these commercially produced timbers which, although significantly cheaper than hardwoods, are far less durable and much more susceptible to insect and fungal attack. The use of hardwoods has, as was mentioned above, implications for the sustainability of forests. Deforestation and poor management has serious implications and the building industry as a significant user of timber materials must therefore adopt a responsible and sustainable approach to its use.

For the majority of timber repairs it should be possible to source a suitably matched material. However, like all conservation work this should follow the rules of minimal repair and honesty so that the

38.9   Timber repairs to a building in Ludlow, Shropshire

historical authenticity of the building is maintained. For example, Figure 38.9 shows timber repairs to a building in Ludlow, Shropshire. The repair is sensitive, but still clearly a modern replacement. Here an earlier timber insert can be seen (aged) at the corner of the second floor just above the bressumer beam and dragon beam at the outer corner edge.

Stone

The geology and abundance of suitable stone in the United Kingdom means that many towns and cities are dominated by stone buildings. Its hard and durable nature means that this has been a favoured building material for centuries. This has varied from the use of field boulders in humble vernacular houses to the use of finely carved ashlar for impressive Classical buildings. Being so heavy, stone was expensive to transport so local sources were sought where possible. Like slate, local geological variations means that traditionally towns varied in appearance depending on the particular stone available in their nearest quarry. However, its durability varies considerably and some stone quarries can produce material which is very vulnerable to weathering whereas others are infinitely more durable.

The geology of the United Kingdom is extremely complex, but the main stone types are summarised as follows:

Sandstone: this is the most predominant building stone in the United Kingdom and is a sedimentary rock which varies greatly in colour and texture depending on its source. It is found throughout central Scotland, eastern Wales, the Midlands and the north of England. It can be finely carved if of suitable type and lends itself to Classical architecture such as in Edinburgh where Craigleith sandstone was widely used and in Bath where Bath stone was the favoured material. However, some sandstones prove to be unsuitable for building purposes because of their tendency to weather poorly.

Limestone: this is a widespread rock formation which is comprised of calcium carbonate and varies from soft chalk to considerably harder carboniferous limestone. Locations are from Dorset in the south to Lancashire in the north. It is widely used for walling materials and is light in colour but varies locally. Chalk is more difficult to use as a building material because it is so soft, but is utilised for some limited building purposes.

Granite: this predominates in northeast Scotland where Aberdeen, known colloquially as the "Granite City" displays a great variety of this extremely hard igneous rock. Rather cold and grey in appearance, it is resistant to spalling and is therefore a very durable if somewhat aesthetically unforgiving building stone.

Flint: flint is 98 per cent pure hydrated silica and is virtually indestructible, but despite this apparent strength it can be easily fractured. In their natural state, flints are generally 5–12 cm, usually rounded by water erosion. In terms of geographical spread, it is found particularly in the chalky deposits of the southeast of England and as a building material is favoured especially in rural Berkshire and also in Norfolk and Suffolk.

As a building material, flint has been used since Roman times but it is a slow method of construction because the mortar must dry out after a section is completed or the wall will bulge out. Construction cannot be undertaken in the winter months as lime cannot reliably be used in the winter season. Building in flint is slow and expensive, and a large quantity of material is often required to fill the interstices between the irregularly shaped stones. However, this will depend on the type of

38.10   An example of modern flintwork in Norfolk

flint used as squared and knapped flint is very tightly jointed. Some types of flint use other materials such as pebbles in coastal areas and galleting, to reduce the void filling.

The strength of the wall is dependent on the nature of the mortar as too wet a mix can create problems and the water content must be just right as flints cannot absorb moisture. However, the use of bricks added in random may have assisted water absorption and improved drying rates, although chalk was the traditional material used for backing flushwork and other knapped work as a means of absorbing moisture.

Flint is a more useful material when combined with other building materials such as brick or stone for quoins and dressings, particularly as pure flint walls require large quantities of material and are labour intensive.

It is possible to date flint buildings through the method of construction. Early buildings dating to the thirteenth century tend to be roughly coursed. This had become neater by the fourteenth century and certainly knapped flints of irregular sizes were being used in superior buildings. "Knap" means to strike or rap and is a method for splitting flints. During the 1320s "flushwork" was used such as at Butley Priory near Oxford. Here, knapped flint is set into dressed stone to form decorative panels and by the end of the century chequers were used.

Into the fifteenth century the flints were more carefully graded and coursed and flush tracery is found in this period. Geographical variations are also evident, such as the use of kidney-shaped flints on the Sussex coast. The decorative uses for flint were further developed during the sixteenth century with the adoption of alternate bands of stone and flint. This approach is both decorative and also functional as it makes the local stone go further and makes for a stronger wall. Flint remained fashionable into Regency and Victorian times when it was used to create polychrome effects, although ultimately it was cheaper to build in brick. The use of flint in several forms is demonstrated in St Edmunds Church, Southwold. Here there are both decorative chequer panels within the stone and emphasis of the tracery designs of the parapet together with its use as a constructional walling material of the tower and nave.

Although the geology of the stone is obviously crucial, the way in which the stone is dressed is also significant. Early use of stone was of undressed rubble, but for prestigious buildings such as cathedrals, stone was carefully sought and cut. The use of stone varies depending on the building type, its prestige and location. It is a highly versatile material and in the hands of a skilled stonemason, magnificent buildings can be the result.

38.11   St Edmunds Church, Southwold, Suffolk

In its humblest form, rubble walls for domestic or farm buildings may be as simple as field boulders sought from the surrounding countryside or stones acquired from small, local quarries. Rubble can be laid randomly or may be brought to courses where it is partially levelled or, finally, it may be regularly coursed. The mortar would generally have been lime but again, would have varied locally and some mortars may include locally sourced clay. There may also be regional variations and introduction of different materials to produce a decorative effect. This may include the use of pinning stones, dressings or bond variations of a different geologic source which vary in texture and colour.

In contrast to the vernacular style of rubble walls, ashlar was reserved for the most important buildings. Thin sections of 'freestone' facing buildings and very fine joints presented a building of distinction.

Many stones, particularly sandstone and limestone, will weather and this may affect the structural integrity of a building. Recent high-profile cases of falling masonry indicate the dangers posed by weathering of stonework and the need for vigilance and maintenance. Not all stones will weather at the same rate and atmospheric pollutants, aspect and location can all affect weathering, as well as the source of the stone and the way it is laid or dressed.

38.12   A flint wall

Stone: conservation implications

Clearly the durability of stone as a walling material is dependent on a number of factors from its basic geology to the way it has been quarried and then used. Subsequent factors such as atmospheric pollution, cleaning of stonework and plant growth all potentially have a significant impact on deterioration of a stone wall. While some deterioration may be due to the inherent nature of the stone or the way it has been quarried, good maintenance can at least help to prevent more rapid deterioration. Allowing gutters to leak can cause staining and erosion and encourage plant growth. Plants then encourage further erosion by encouraging water to collect and the roots of certain plants like ivy, can be particularly damaging.

The closure of quarries both local and larger means that it is very difficult to obtain stones of an exact match for repair of existing historic buildings. The use of stone which is not a good match can cause serious deterioration in stonework, similar to the use of hard cement renders. It is therefore very important that the source of the original stone is understood and fully investigated before any intervention works are undertaken.

38.13   A rubble wall

A further implication for conservation is the lack of available skilled stonemasons to carry out specialist work. The continued deterioration of buildings, as evidenced by recent deaths and injuries from falling masonry, are indicative of the serious nature of the problem. There are significant cost implications in the maintenance of substantial stone buildings. It also needs to be remembered that particular building types and materials may require specialist local skills which are not always available. The lack of suitable expertise therefore has considerable implications for repairs, for example, of flint buildings which require tradesmen who are familiar in how to use this material. With changes in employment opportunities and in apprenticeships, the skills base needed in many cases is inadequate.

Re-pointing of stone buildings can, if improperly undertaken, lead to a diminution of significance. Careful analysis of the stone and its weathering qualities and original pointing methodology must be carried out prior to repair or intervention work. Incorrect method and material usage can seriously damage historic buildings and lead to loss of visual importance. In Figure 38.18 the poorly executed re-pointing and inappropriate use of materials is evident in this early sandstone house in Staffordshire. The method of re-pointing has completely overtaken the original stone texture and appearance and is dominating the façade. A plastic door and windows, along with modern roofing material, have detracted from the building's significance, originality, value and authenticity. The absence of a visual lintol over the front

38.14   Ashlar wall

38.15   Deterioration of stone carving, Exeter Cathedral

38.16   Growth of buddleia in a stone wall

door is a classic example of inappropriate, untrained, unsympathetic, damaging intervention.

The implications of the fashion in the latter twentieth century for cleaning of stone buildings is still not yet fully understood. This was particularly popular during the 1960s and 1970s in order to improve the appearance of soot- and grime-covered buildings, some of which had become virtually black instead of their original buff colour.

While it was believed to be beneficial both aesthetically and for the stone itself, it is now proving that such cleaning methods were in fact

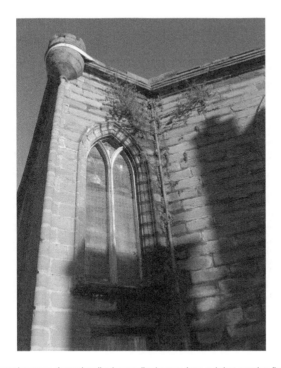

38.17   Poor maintenance of guttering allowing a wall to become damp and plant growth to flourish, potentially damaging the stonework

potentially very detrimental to the integrity of the stonework. These methods included both abrasive sand-blasting and the use of chemicals to remove the soiled layer. As little or no prior research had been undertaken these works were uninformed in terms of the potential damage to the stonework.

The implications from the legacy of this work are that the surface of the stone may have become pitted and damaged, together with a loss of architectural detailing through the use of heavy-handed techniques. The result of cleaning can therefore be an initial improvement in the appearance of the building but, in the long term, the stone is potentially more vulnerable to deterioration and growth of undesirable materials like algae. On-going research in this area is a vital source of information to establish the long-term implications of this work.

38.18   Poor pointing of stone, Staffordshire

Brick

Bricks have been used over about 3,000 years, from their early origins as mud-based, air-dried "buns" which were similar in size to present-day bricks to great bricks which were devised to reduce the impact of brick tax. Bricks have survived periods of being less fashionable such as in the early nineteenth century when they were covered over with stucco, becoming popular again due to a revival during the Victorian period. The enduring qualities of bricks are demonstrated by continual use over a significant period of time with periods of reduced interest but a general continuum of use.

A brick is a clay-based unit which does not exceed a length of 337.5 mm, a width of 225 mm, or a height of 112.5 mm (Bailey and Hancock, 1990). The requirements for brick manufacture are two-fold:

1   use of the correct material
2   the necessary level of skill.

In terms of the correct material, two types of clay are needed; one plastic and one sandy to stop the plastic clay from shrinking and warping. In historical brick production, these were sent through the pug mill to even out the texture, pressed into wooden moulds and left to dry for a minimum of two weeks (dependent on the season) before being fired in a kiln for at least 48 hours. Clay was dug locally and there was no standardisation. Brickwork is a very durable building material if properly made, but defects may arise either due to using the wrong clays or through defective manufacture methods.

Historically, the Romans used bricks for arches and vaults but primarily for walling. However, it was the fifteenth century before the word "brick" was used widely, the items having previously been known as wall tiles. One of the earliest examples of post-Roman brick building can be found in the arches of a church at Polsted in Suffolk which was built around 1160. During the thirteenth and fourteenth centuries, Flemish bricks were imported where there were strong trading links with the east coast of the country. Ultimately, immigrant Flemish brickmakers produced bricks made locally in this country. These were generally nine inches long by four inches wide and two inches high, which is similar to a standard modern brick. Surviving early examples of brickwork dating to the fifteenth century include Tattershall Castle in Lincolnshire (1434-c.1448) and Hurstmonceaux Castle in Sussex (1440) which is now occupied by the Royal Observatory.

Brick really blossomed at the start of the sixteenth century, especially after being used in the great palaces of Cardinal Wolsey such as Hampton Court and, more importantly, Henry VIII. By the end of that century, even the most middling yeoman houses had at least a brick – and therefore fireproof – chimney. During the eighteenth century most ordinary houses in England were constructed of this cheap and readily available material and it was even used to front timber-framed houses to make them more fashionable. However, where there was local stone readily available or an absence of clay, brick was much slower to be introduced. Classical architects tended to imitate stone if they had to use brick and bricks were also used for stable blocks of large country houses.

In 1784 a Brick Tax was introduced. It was subsequently increased in 1794 and 1803. In the country this encouraged the use of wood, but it does not seem to have affected urban areas as brick was still cheaper than stone. However, there was an increase in brick size to a thickness of three inches, up from 2.5 inches, as clearly less bricks at the larger size were needed in order to create the same wall. During the Regency period of the early nineteenth century there was a strong fashion for hiding the brickwork under a coating of stucco. This resulted in on-going maintenance problems in trying to maintain its appearance through repainting and repairs. However, in 1850 the Brick Tax was removed and the fashion for stucco faded.

The localised nature of brickworks meant that bricks were regionally distinctive. Clifton-Taylor (1987: 223) highlights these variations, stating:

Thus up to the Industrial Revolution brick, no less surely than stone, exhibited its own local characteristics, in harmony with the land upon which it stood and out of which it came. It has been well said that at one time there were as many different varieties of brick in England as there were of homespun loaves.

However, during the mid-nineteenth century, the mass production of bricks produced some colours which lacked aesthetic sensitivity and no longer had that local connection. At this time brick manufacture became mechanised and clays previously thought unusable in the north Midlands and north of England now became available. From 1880 the immense bed of Oxford Clay at Fletton was exploited and by 1889 was producing 156,000 bricks per day. The development of the Hoffman Kiln, able to burn continuously, helped this mass production. The use of the now extensive canal and rail system meant that bricks could be transported all over the country from these massive brickworks. Clifton-Taylor (1987: 229) suggests that the development of mass-produced bricks was artistically "calamitous" and because the small local brickworks were not able to compete with the large ones this had a direct impact on brick design. He states: "The distinctive character and colour harmonies of particular localities were completely disregarded: these smooth, insensitive bricks even found their way into stone districts in which visually they had no aesthetic justification whatsoever."

As the northern mill and mining towns rapidly expanded to accommodate thousands of workers, mass-produced houses were erected using brick as the primary construction material. These terraces of two-storey cottages are found throughout Lancashire, Yorkshire and Durham. Clifton-Taylor (1987) describes these as "so hideous that it almost seems as if lack of visual appeal had been deliberately

38.19   A Tudor country house – Blickling Hall, Norfolk

sought", and in relation to the many northern towns he suggests that "cheapness now dictated that the building material should nearly always be brick, the effect of which is altogether lamentable". The use of unskilled labour and the adoption of poor-quality materials such as poorly produced mortar using reduced lime content produced friable mortar and unstable walls.

In contrast to the domination of brick south of the border, it was not extensively used in Scotland as a building material until the twentieth century when it had become the favoured material for modern housing construction. The plentiful supply of stone made Scotland very much a country of stone with a few notable exceptions, such as the Carse of Gowrie, Perthshire, where Errol brickworks still operates today.

Overall, the success of brick as a building material was a matter of economics as it was better value than any other product:

- costs were less than quarrying and transporting stone;
- skilled stonemasons were higher paid than brickmakers;
- they are convenient to handle and more or less uniform in size which makes construction costs cheaper;
- well-burnt bricks are extremely durable and hardly deteriorate with polluted air;
- brick is non-combustible;
- brick is a warm material: being porous it is not a good conductor so is warm in winter and cool in summer.

Brick types, colours and textures

The method of manufacture and the types of clay used will influence the colour and durability of the brick. There are three main categories of brick: commons, facing and engineering. Common bricks will be used in areas where their aesthetic appearance is not important, such as for basements or internal walls which will then have a plastered surface. Facing bricks are, however, designed to be seen and will be used to face buildings. Engineering bricks are specially manufactured in terms of their strength and lack of porosity in order to be suitable for particular purposes such as bridges or sewers.

For appearance, the colour of a brick is of particular importance. This is influenced by two main factors:

- **Clay**: this is the main determinant of colour, together with the impurities which act chemically as staining agents: iron, manganese, cobalt, lime and sand. Clays from similar geological formations tend to burn to the same colour. For example, the marls of the Triassic produced the brightest reds. The shales of the northern coal measures produced "Accrington bloods" which are extremely red and very hard. Yellow brick types such as gaults are produced using high chalk content clays like those round Cambridge. The fashion in London for yellow-coloured bricks during the eighteenth century was facilitated by adding finely ground chalk to the brick earths. This was made easy for the brickmakers because there were large chalk deposits underlying the excavated brick clays of the London Basin. This fashion for yellow colour changed the face of London architecture during the eighteenth century (Woodforde, 1976). This takes up the Palladian ideal of buff/yellow being the only suitable colour for a brick house.

- **Firing**: if the clays were not properly mixed by the pug mill then different lumps could burn to different shades. The position within the kiln is also important as the bricks exposed to the greatest heat would emerge from the kiln the darkest. The character of the fuel might also affect the colour – some woods contain potash which affects the glazing. However, modern kilns control colour by controlling the amount of air admitted in the later firing stages.

Local differences were often accidental because the temperature of the kiln could not be accurately controlled and the chemical composition of the clay was not known. These variations led to a richness in the brick architecture of the Tudor and Stuart period.

Historically, prior to the eighteenth century, brick buildings of any colour other than red were uncommon, although there were many shades of red. The browner or redder the brick the greater the iron content of the clay. However, other colours did evolve. The main brick colours found are as follows:

White bricks: these have a high lime content. Found in Suffolk, Norfolk, Bedfordshire and Peterborough.

Brown bricks: these occur widely, especially in the Vale of York. Also London "stocks" are yellow-browns, a high-grade brick moulded on a stock, a wooden board with a small frame or mould to contain the clay. These were produced from the large supply of brick clay in the Thames valley.

Yellow bricks: use of yellow bricks may be coincidental with regions that traded with Holland where small hard bricks known as "Dutch clinkers" came from ship ballast.

Grey bricks: these are found in south Oxfordshire and Berkshire, from Thame to Hungerford. Many were made in the Reading to Newbury area. For example, Donnington Grove, a Gothic house of the 1760s near Newbury, is entirely faced with grey bricks. Grey was held in such high esteem in the latter eighteenth century that where lime was lacking, red bricks might be given a thin "skin" of grey by adding salt to the coating of sand before firing.

Blue bricks: these are of comparatively recent origin from the durable clays of the coal measures and very hard and damp resistant so they make good engineering bricks, e.g. Staffordshire blues.

During the Victorian period, polychrome brickwork became fashionable. This is where different coloured bricks are used in a decorative or geometric design (Figure 38.20). Reading in Berkshire is one of the best concentrated examples of this use of brickwork.

A brick wall is an aggregate of small effects which does not have the impressiveness or splendour of stone. However, using particular materials and bonds, a brick wall could still be aesthetically very pleasing. The appearance of brickwork is not only influenced by the colour of the bricks themselves but will be materially affected by the form of the bond used. Bonds developed over a great period of time and were characterised by the proportion of headers (ends of bricks) to stretchers (faces of bricks). Some examples of the various types of bonds are illustrated in Figures 38.21–38.24.

The best brick textures depend on the use of sand and are contrived. Sand was either dusted on the unfired lump of clay or dusted into the mould or both. Fine sand was used in late Stuart and early Georgian to give a smooth finish.

38.20   Victorian polychrome brickwork, Norwich

English bond

Flemish bond

38.21   English and Flemish bonds

English garden - wall bond

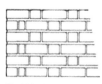

Flemish garden - wall bond

38.22   English and Flemish garden wall bond

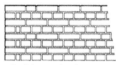

English cross bond

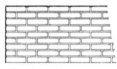

Stretcher bond

38.23   English cross bond and stretcher bond

'Soft' bricks are produced by adding more sand and baking, not firing. They are used for dressings – pilasters and window surrounds – or for gauged work which involved cutting and rubbing. Soft bricks were needed for voussoirs above windows and could be cut with a special saw to obtain the required shape before being rubbed down to make the finest joint.

Other brick appearance approaches were achieved through the use of brick tiles. During the eighteenth century, tiles which imitated bricks, known as mathematical tiles, were used to re-face buildings and had the added advantage of avoiding the 1784 Brick Tax. They were particularly popular in Sussex and Kent, especially in the Sussex town of Lewes. Similarly, tile-hung walls were found in particular English counties, notably Kent, Sussex and Surrey, together with parts of Hampshire and Berkshire.

Brick: conservation implications

A good understanding of how original bricks were made, their composition and degree of hardness will all materially affect conservation work. The type and composition of the mortar used to build brickwork is as equally important (in some cases more so) as knowledge of the bricks themselves because adopting the wrong mortar for repair and re-pointing may cause catastrophic damage to the wall. It is essential when undertaking work to historic brickwork that the original composition of the mortar, both in respect of the lime used and in respect of the sand used, is carefully analysed and replicated. The type of lime and type and grading of sand must be carefully studied and copied in intervention work if the possibility of damage is to be avoided. The use of cement mortars in historic brickwork must not be considered as its use could lead to irreparable damage. The colour of the sand used in the mortar will have a dramatic effect on the overall appearance of the wall, as will the method of pointing. Mechanical removal of old and failed mortar may result in damage to the arrises of the bricks and will result in a widening of the finished joint width, marring the appearance of finished work. It is therefore advisable to undertake the cleaning out of joints by hand methods. This is a painstaking but essential process to ensure that the work does not detract from the finished appearance of a repaired section of brickwork.

Where possible, replacement bricks should be either sourced from specialist suppliers who can replicate the original brick and firing conditions or be sourced from local buildings that may have been demolished but made use of the same bricks. The latter method of sourcing has implications in conservation terms in that the bricks

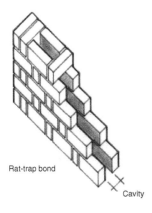

Rat-trap bond

Cavity

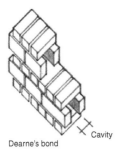

Dearne's bond

Cavity

38.24   Rat-trap bond and Dearne's bond

38.25   Mathematical tiled façade – Norwich – a telltale being the jetty at first-floor level

recovered will have been generated by loss. However, in some cases the use of salvaged materials may be the only option.

Architectural ceramics: terracotta and faience

Terracotta and faience are defined by Ashurst and Ashurst (1988: 66) as "moulded clay products made from fine, pure clays, mixed with other materials such as sand and pulverised fired clay. They are usually well vitrified and have a hardness, compactness and sharpness of detail not normally obtained in brick."

These materials found particular favour from the mid-nineteenth century as mass-production techniques were developed. They remained popular through the Edwardian era and into the inter-war period.

While faience is usually associated with a glazed finish, terracotta is generally unglazed, although it is possible to glaze terracotta. However, the glazing is usually transparent so that the colour of the clay is evident, whereas with faience the colour of the glaze, such as white, is the dominant colour (Pearson, 2005).

Like bricks, terracotta and faience will vary depending on manufacturing methods, clays and firing. However, there were certain proprietary types made, one of the most well known being Coade stone which Ashurst and Ashurst (1988: 70) describe as "a uniformly vitrified and highly durable material". This was made in Lambeth by Eleanor Coade from 1769, and although clearly not stone, imitated its appearance. It was used to make architectural ornaments such as sculptures and garden ornaments.

The other leading firm in the production of terracotta and faience were Doulton & Co. Ltd of Lambeth. They produced various products from the 1820s, including Doultonware (introduced in the 1870s) and Carraraware (from the 1880s). Examples include the terracotta Doulton Fountain at Glasgow Green, Glasgow (1887–1888) for the Glasgow Exhibition of 1888 and its nearby neighbour, Templeton's Carpet Factory (1888) designed by William Leiper and constructed in faience, polychrome brickwork and terracotta, based on Venice's Doge's Palace (Pearson, 2005) (Figures 38.26 and 38.27). In London the Victoria and Albert Museum in South Kensington makes wonderful use of terracotta but the building also contains "some of the finest products of British nineteenth century ceramic design" (Pearson, 2005: 224). The nearby Natural

38.26   Doulton Fountain, Glasgow Green, Glasgow

38.27   Templeton's Carpet Factory, Glasgow

History Museum (1873–1881) was also "the first major public building in Britain to be built with an entire façade of terracotta, the scale and richness of its decoration was unmatched" (Pearson, 2005: 226).

During the inter-war period the popularity for smooth and streamlined buildings made faience a natural choice. The material was particularly well suited for the designs of the Art Deco and Moderne movements and companies like Burtons (the tailors) used it for many of their purpose-built retail stores during the 1930s. White faience with stepped parapets was widely adopted and although the ground floors have often been substantially altered in recent years, the faience on the upper floors is still clearly evident.

Conservation implications: terracotta and faience

Architectural ceramics are complex materials which can be subject to decay and deterioration. While this may be caused by original manufacturing defects such as inadequate firing, other problems can arise through salt crystallisation and water ingress. The fixing systems can also fail or there may have been an inappropriate approach to repairs. A lack of good repair maintenance such as removal of vegetation can also cause deterioration in glazes.

Any damage must be carefully inspected as its extent may not be obvious. Ashurst and Ashurst (1988: 72) state that "particularly with terracotta, the outward signs of decay do not always indicate the more serious problems within" and therefore close inspection, possibly using an endoscope and a corrosion meter may be required.

While the numbers of companies manufacturing terracotta and similar products are limited, there are companies that can carry out suitable specialist repairs and reproduction of materials where appropriate.

Iron

Iron has been used since the "Iron Age" of 750 BC (Swailes, 2006) and was certainly used in Roman times. Although wrought iron was widely used in this country, it was from the latter eighteenth century that cast iron became the more popular material due to the invention of coke smelting and re-melting furnaces (Hume, 1992). Ashurst (1998: 15) states:

> the breakthrough occurred in 1794 with Wilkinson's invention of the cupola, a small blast furnace used for remelting pig iron rather than smelting ore. It provided cast iron by a quick, relatively economical, simple process and was therefore the impetus for the establishment of a multitude of small foundries.

Early uses for cast iron included grave slabs, firebacks, cannons and cooking utensils, but by the nineteenth century it was used for "every conceivable piece of equipment for buildings" (Gloag and Bridgewater, 1948: 278). Bannister (1950) says that from the 1780s cast-iron columns were used in churches such as St Annes, Liverpool (1770–1772), but the first attempt at a truly fire-resistant building utilising cast-iron columns was the Calico Mill in Derbyshire by William Strutt, built 1792–1793.

Cast iron had wide appeal and was a material used for both structural work such as columns in cotton mills like Stanley Mills, Perth, or for decorative ironwork such as railings and balconies in Edinburgh's New Town. This is due to its various qualities as outlined by Ballantine (1997):

- has the ability to be melted or poured into different shapes;
- has a long life-span because it is very resistant to wear and corrosion;
- it has poor tensile strength but high compressive strength;
- it conducts heat well so can be used for heating and cooking.

Cast iron has its limitations, notably it is weak in tension but very strong in compression. Its limitation therefore is that it cannot effectively be used in tension structures. It cannot be bent, being weak in tension, and cannot be forged or worked into shapes and so it relies on its ability to be cast. With these limitations its use structurally was relatively limited to compression members such as posts and columns and decorative elements of construction. It was not until 1783 when Henry Cort developed a process known as "puddling" that the more versatile wrought iron became available in quantity (Koppelkamm, 1981).

Puddling adopts the principle of stirring air into the melted iron, thus abstracting carbon and producing malleable or wrought iron in quantity. This material was particularly useful in facilitating structures where tensile stresses had to be resisted, such as iron bridges. The most famous of these structures at Ironbridge in Coalbrookdale in Shropshire was built to cross the River Severn. The bridge is a cast-iron structure which was built in 1779 by Thomas Pritchard (architect) and Abraham Darby (engineer). Coalbrookdale played a central role in the Industrial Revolution and is now designated as a UNESCO World Heritage site.

38.28   Iron bridge at Coalbrookdale

Cast iron proved to be a highly popular architectural material during the nineteenth century because of its numerous advantages which Gloag and Bridgewater (1948: 96) outline as being "its fire-resistant qualities, its cheapness, simplicity of manufacture and resistance to heavy loads". Its versatility and low cost made it an attractive material. It was also possible to produce quite decorative work by creating patterns which were carved in wood and closed moulds which were used for complex articles (Smith, 1985). Ballantine (1997) states that in a foundry the greatest skill is in pattern-making, which requires skills from a sculptor and a master joiner. Smith (1985: 63) confirms this saying: "The skill of the pattern-maker working in conjunction with the moulder was crucial in solving the difficulties of pattern assembly and withdrawal."

Scotland proved to be a major centre of iron-founding and in 1880 there were over 6,000 people employed in ironworking along the Forth and Clyde area (Gloag and Bridgewater, 1948). These included such famous manufacturers as Walter MacFarlane (Saracen Foundry), George Smith (Sun Foundry) and the Carron Ironworks, Falkirk, established in 1760. The existence of so many foundries was to have a significant impact on architecture as the product developed and was experimented with as the new foundries grew. Gloag and Bridgewater (1948: 310) state:

> Apart from the increased use of the material for fittings and equipment, new large scale uses were tried out. Windows, panels and apron pieces, shop fronts, and indeed whole facades of buildings, were successfully cast and were used in Britain, and exported to all parts of the world.

Glasgow famously adopted the material and used it to great effect in the construction of a number of cast-iron warehouses. The first of these were built in the 1830s and several were constructed throughout the Victorian period, including the Ca' d'Oro building and Gardiners Warehouse, Jamaica Street. Only a handful still survive, but these impressive buildings led the way in iron construction. The construction of so many of these in one location reflected the construction of iron-framed buildings in some American cities.

Wrought and cast iron, and later steel, were fundamental materials used during the Industrial Revolution and in early Victorian structures.

The development of steel was facilitated by the invention of the Bessemer converter, invented by Henry Bessemer in 1856. The Bessemer converter turned iron into steel by blowing air under pressure through the molten iron and thus producing a metal with even less carbon content than that permitted by the former puddling process to create cast iron. The metal thus produced was exceptionally malleable and could be easily forged or pressed and rolled into various shapes – for example, solid steel beams could be used as opposed to composite, riveted steel beams and sections of wrought iron. This paved the way for even greater developments in steel-framed structures than had previously been facilitated by wrought and cast iron.

The development of these metals permitted the development of the railway systems in the nineteenth century which then played a key role in the economic expansion of Britain. Many of the United Kingdom's main-line railway terminals use both wrought and cast iron in their long-spanning roof structures and numerous examples have decorative and attractive ironwork.

Cast iron also found great favour in the early-to-mid-nineteenth century for the construction of conservatories, palm houses and exhibition buildings, the greatest being Joseph Paxton's Crystal Palace for the Great Exhibition of 1851. Iron was ideally suited for the creation of elegant winter gardens and glass houses, popular during the Victorian period through a development in botany and the availability of exotic plants. Examples are found at Kew, Glasgow, Edinburgh, Chatsworth House, Derbyshire, Syon House and, amongst others, the enigmatic Palm House structure at Bicton Gardens in Devon with its gossamer-like roof tracery.

38.29 People's Palace, Glasgow Green, Glasgow

38.30 Kibble Palace, Glasgow

Iron: conservation implications

Iron remains a widespread material in Britain with its use as a decorative and structural material. During the Second World War people were encouraged to remove their iron railings so in many parts of the country there is limited surviving ironwork and the urban landscape is the poorer for this. However, in some more remote locations where this removal did not occur, such as Stornoway and Shetland, good examples of what our streets once looked like do survive. In some cases reinstatement has taken place and the use of ironwork is still seen extensively in the Classical cities of Bath and Edinburgh.

Like all materials, cast iron requires appropriate maintenance or it will deteriorate. It is not difficult to find examples where there has been significant damage and resultant loss of fabric. A well-maintained paint finish is the best method of preventing deterioration.

The closure of the majority of the major iron foundries means that it is much more of a specialist product and is no longer produced on a mass-production scale in the United Kingdom. It may therefore be expensive to replicate items. However, there are now a number of small foundries operating which will carry out such work (see www.scottishironwork.org for more information).

Glass

Glass was used by the Egyptians in the second millennium BC, but it was the Romans who first used it for architectural purposes (Weaver, 1993). It is believed to have been introduced to England by the Romans, although it is said that French glassmakers made the windows of a church in Wearmouth in AD 675, suggesting that the craft was forgotten when the Romans departed (Morris, 2001). French glassmakers settled in the Wealden area in 1226 but throughout the medieval period it remained prohibitively expensive and so was restricted to high-status buildings such as churches. It also appears that the quality of English glass was considered to be inferior to that of the Continent and, as a consequence, large quantities were imported from France, particularly Burgundy, Lorraine and Normandy (McGrath and Frost, 1961). By the mid-thirteenth century most churches would have had some or all of their windows in glass, but for domestic properties it remained exclusive to the king and the most important members of the royal household.

Crown glass was said to have originated around 1330 by Philippe de Caqueray at Le Haye near Rouen. This type of glass predominated in Normandy. However, McGrath and Frost (1961) suggest that the principle seems to have been used by the Syrians in the first three centuries AD and that the great Venetian glassmakers also learned from the Syrians how to spin their *rulli*. He says that it is possible that the Normandy crown glass was the result of European contacts with Syria in the thirteenth century. From the mid-sixteenth century glass-making techniques were improved, particularly through the influence of foreign glassmakers from France and Venice.

Crown glass was made by spinning a globe of glass with the result that the centre had a bullion or "bull's eye". It was slightly curved and was partly blemished with striations, but was very bright and lustrous. McGrath and Frost (1961) state that by the end of the eighteenth century, panes of crown glass were being made 24 inches by 15 inches, and its popularity for windows grew as broad glass popularity receded.

A further important development in glass manufacture was the invention of plate glass by the French who developed a technique of casting or *coulage* from 1665 onwards. However, it was 1773 before cast plate glass was manufactured in England. The British Plate Glass Company established its works at Ravenshead, St Helens, but it was extremely expensive and tended only to be used for mirrors and coaches, and according to Morris (2001) it was not widely adopted until the early nineteenth century.

The adoption of glass was partly hindered not by the limits of production techniques but by the various excise duties which applied to it. This heavy taxation burden affected development and sale of glass and ultimately its use in architecture. McGrath and Frost (1961) quote *The Plate Glass Book* published in 1758 which gave the price of a silvered plate 60 inches by 42 inches as £81.17.0, of which £37.10.0 represented excise duty.

Window Tax was introduced in England in 1696 and levied according to the number of openings on all inhabited houses worth more than £5 per annum and with six or more windows. It was levied dependent on the number of openings and the result of this was that some windows were blocked up to ameliorate the impact of the tax. Under the Union of 1707 the tax was extended to Scotland, but according to McWilliam (1975) this had less impact than the glass taxes which were heavy excise duties introduced in 1746. Eldridge (1958) confirms that these were a "serious restraint" on the use of plate glass.

There were three duties which were payable on the glass-making industry:

- an annual payment for each glasshouse for a licence to manufacture glass;
- a payment per pound on all glass melted in the pots and ready for use;
- a payment per pound on the excess in weight of manufactured glass over 40 per cent (later 50 per cent) of the calculated weight of molten glass.

As the tax was by weight, not size, the Dumbarton glassworks which created larger clear glass but which was thicker could not compete with the crown glass from Stourbridge which was effectively lighter (McWilliam, 1975). It was 1845 before the taxes were repealed, but the removal of import taxes at the same time meant the demise of many small glassworks which had depended on this protection for survival. The cost of glass plummeted with the removal of the duty. McGrath and Frost (1961) quote that from 1827 to 1844 the average for fourths crown glass tables was 6s 1d per table, but by 1849 this had dropped to about 2s each.

However, the existence of the duties did not prevent the Chance brothers from creating broad or sheet glass by a new technique known as cylinder glass in 1832. Robert Chance had perfected a technique already used on the Continent and by 1851, with the removal of excise duties, the Chance brothers supplied almost one million square feet of glass for the Crystal Palace.

The development of glass had a major impact on window design and its development. Early windows were side-hung and leaded and diamond-shaped panes of small glass were used. It was the late seventeenth century before the sliding sash was first used in any quantity at Chatsworth House in 1676–1680. Their use in the royal palaces of Windsor Castle, Kensington Palace and Hampton Court increased their popularity and fashionability, and they lent themselves to the Classical architecture and proportions of the eighteenth century.

Specialist glasses

There are many specialist techniques in glass production which allow glass to be not only a practical but also a highly decorative material. Stained and painted glasses have been used to great effect for centuries in church buildings. This material was also popularised during the late nineteenth century and stained glass was extended in use to both residential and commercial buildings, many designs inspired by the Art Nouveau period.

Some very specialist materials were developed to resolve particular lighting problems. Prismatic glass was developed in order to light basements or deep buildings at a time when electric light was not available. This glass uses the concept of bending daylight so that it passes more horizontally through a building. The glass blocks are smooth on one side but have a series of prisms running across the inside face of the glass which directs the light in a predictable way. Prismatic glass is believed to have been made in the eighteenth century, although it was the nineteenth century before it became more widely used. Neumann (1995: 89) describes prismatic glass as:

> characterized by small horizontal triangular ribs on the interior face that refract light rays deep into a room. Commonly used in pre-electric commercial buildings to improve lighting, this type of glass was originally produced in tile form and later in larger sheets.

Prismatic glass was popularly made as glass blocks which may still be seen in the stallrisers of shops (the section below the windows) and in other commercial buildings to light basement areas.

Fashions for smooth, clean lines such as those in the 1930s encouraged the use of glass products such as Vitrolite. This rolled opal glass was manufactured by the Pilkington brothers and was available in a variety of colours, although black, cream and yellow were the most commonly used. In the inter-war period it was used for most of the glass-faced buildings of the 1930s and was widely used for fascias and shopfittings and for wall lining in interior work. Because rolled opal is impervious to so many different substances such as blood, methyl violet and various acids it is very useful where hygiene is required such as in hospitals or kitchens.

Glass: conservation implications

Historic glazing is of considerable interest and significance. Small blemishes and the rippled effect of earlier glass adds to the character of traditional buildings. The design of sash windows is also aesthetically pleasing and the replacement of these with modern alternatives is rarely successful. However, as a fragile material, glass can be broken and it may be necessary to seek suitable alternatives by companies who manufacture traditional glasses. However, getting replacements for specialist glasses such as Vitrolite, which is no longer manufactured, can be a problem where such material is in need of repair. There may also be a lack of understanding of how to carry out such repair techniques and in terms of the particular way that such a material was used.

The increasing demand for double-glazed windows in both domestic and commercial buildings, often with the use of upvc frames, can seriously detract from the aesthetic value of a historic building. There are also uncertainties about the durability and environmental sustainability of materials like upvc, and their heavy frames are generally found to be unsuitable for historic buildings, even where they are designed in a sash format. Where possible, historic glass and the surrounding frames should be retained, particularly as many sash and case windows are capable of renovation and continued use.

Work to stained and leaded glass is highly specialised and requires the attention of appropriately qualified conservation experts.

Chapter **39**

Conclusion

This part has set out the main building materials to be found within the historic buildings of the United Kingdom. As with architectural styles, materials have altered and adapted over time as manufacturing methods and sources of materials have changed. Some materials have been highly fashionable but short-lived, such as Vitrolite, but others have an enduring and long-lasting place in construction, such as brick. However, all of these must be fully understood prior to any intervention work being undertaken. Some will require specialist knowledge due to their particular composition and it is always advisable to investigate this fully to ensure that the correct approach to repairs is undertaken.

Recommended reading

Williams-Ellis, C. (1999). *Building in Cob, Pisé and Stabilised Earth.* Shaftesbury, Donhead Publishing.

Walker, B. and C. McGregor (1996). *Earth Structures and Construction in Scotland.* Edinburgh, Historic Scotland.

Keefe, L. (1993). *The Cob Buildings of Devon 2: Repair and Maintenance.* Devon Historic Buildings Trust. Online, available at: www.devonearthbuilding.com/leaflets/the_cob_buildings_of_devon _2.pdf.

Houben, H. and H. Guillard (1989). *Earth Construction: A Comprehensive Guide.* London, Intermediate Technology Publications.

Norton, J. (1997). *Building with Earth: A Handbook.* ITDG Publishing.

Pearson, G. (1992). *Conservation of Clay and Chalk Buildings.* London, Donhead Publishing.

Cob Paper (n.d.) Online, available at: www.ihbc.org.uk/Cob_Paper.

Bouwens, D. (1997). *Earth Buildings and their Repair.* Online, available at www.buildingconservation.com/articles.earth/earth.

Woodforde, J. (1976). *Bricks to Build a House.* London, Routledge.

Lynch, G. (1994). *Brickwork: History, Technology and Practice, Volumes 1 and 2.* Shaftesbury, Donhead Publishing.

Bailey, H. and D. Hancock (1990). *Brickwork and Associated Studies Vol 2.* London, Macmillan.

Brick Development Association (2005). *BDA Guide to Successful Brickwork.* Windsor, Brick Development Association.

Brunskill, R. W. (1990). *Brick Building in Britain.* London, Victor Gollancz in association with Peter Crawley.

Brunskill, R. W. (1982). *Illustrated Handbook of Vernacular Architecture.* London, Faber & Faber.

Lynch, G. (1994). *Brickwork*, volumes 1 and 2. London, Donhead.

Lloyd, N. (1983). *A History of English Brickwork.* Montgomery, Antique Collectors' Club.

Clifton-Taylor, A. (1987). *The Pattern of English Building.* London, Faber & Faber.

Wight, J. (1972). *Brick Building in England from the Middle Ages to 1550.* London, Baker.

Part 6

Conservation philosophy, historical context and legislation

Part **6**

Introduction

This part outlines the historical development of the conservation movement and its philosophies, together with a consideration of international charters and current legislation. However, it must be borne in mind that not only does legislation vary from country to country, even within the United Kingdom, but it is also being continually changed and updated as part of a continuum of development as attitudes and philosophy changes. It is therefore the duty of professionals working in this field to keep up to date with changes. As such, this book can only offer an outline of what can be a complex and detailed subject. The reader will be directed to appropriate additional reading material.

Chapter **40**

Conservation or preservation?

At the outset we need to consider what the term conservation means, in particular, what the difference is between conservation and preservation. An examination of definitions and terms can help to explain this.

The ICOMOS Education and Training Guidelines state:

> The object of conservation is to prolong the life of cultural heritage and, if possible, to clarify the artistic and historical messages therein without the loss of authenticity and meaning. Conservation is a cultural, artistic, technical and craft activity based on humanistic and scientific studies and systematic research.

This definition covers not only the primary aim of conservation in ensuring that the heritage is maintained for the future, but that the authenticity and meaning of the sites is also respected. These elements are also reflected in the BS 7913 (*The Principles of the Conservation of Historic Buildings*) which defines conservation as: "Action to secure the survival or preservation of buildings, cultural artefacts, natural resources, energy or any other thing of acknowledged value for the future."

Preservation on the other hand, suggests a 'pickling in aspic' approach to buildings and does not allow for the fact that effectively buildings and their uses change. Fielden (in Earl, 1997) states that conservation is "very largely the art of controlling [or managing] change". Change is inevitable with buildings which have, in some cases, been standing for centuries. Effectively managing change without compromise to the fabric is the challenge of conservation. Change is inevitable if the heritage is to survive at all, so Fielden's modified quotation is particularly apposite in defining that change, although inevitable, must be both appropriate and managed sympathetically.

While there may be no definitive definition of the term conservation, the need to recognise cultural significance and maintain historic structures through appropriate repair, maintenance and where required, re-use, are core elements. Conservation does not necessarily mean ossification and may therefore be distinguished from preservation.

Chapter **41**

Historical context: the nineteenth-century anti-scrape movement

The origins of the modern conservation movement are in the "anti-scrape" movement of the late nineteenth century. The principles of conservative repair were established by William Morris and the Society for the Protection of Ancient Monuments (SPAB) but the philosophical debate had its origins earlier in the nineteenth century.

Early concerns about the over-restoration of buildings were raised by the work of architect James Wyatt (1746–1813). He was given the nickname "The Destroyer" for his work on Lichfield, Hereford, Salisbury and Durham cathedrals at the end of the eighteenth century. Despite the fact that many of England's cathedrals were decaying and in need of intervention, his work was later condemned by the Society of Antiquities. Two other figures that became influential in the conservation debate, Augustus Pugin and John Ruskin were also critical of Wyatt's work as being insensitive and unnecessarily destructive. Wyatt's work involved the subjective re-design of these medieval cathedrals to a subjective, 'idealised' appearance that had never previously existed.

The visionary ideas of the influential writer and evangelist, John Ruskin (1819–1900) were established in his books, *The Seven Lamps of Architecture* (1849) and *The Stones of Venice* (1851). These became cornerstones in the Arts and Crafts Movement of the late nineteenth century. He believed in the value of creative work and rejected Victorian designs which required machine production, thereby destroying the creativity of human labour (Adams, 1987). Like Pugin, Ruskin was interested in constructional honesty. In "The Lamp of Truth" within his book *The Seven Lamps of Architecture* (Ruskin, 1849) he states:

The violations of truth, which dishonour poetry and painting, are thus for the most part confined to the treatment of their subjects. But in architecture another and a less subtle, more contemptible, violation of truth is possible; a direct falsity of assertion respecting the nature of material, or the quantity of labour.

Ruskin suggested three "architectural deceits":

- suggesting a mode of structure other than a true one;
- painting surfaces to represent a material other than what they actually are;
- the use of cast or machine-made ornaments.

In contrast to Ruskin's belief in constructional honesty, Eugène-Emanuel Viollet-le-Duc (1814–1879), a French architect and archaeologist, aimed to recover an ideal of a building. He was appointed by the French Historical Monuments Commission to oversee works to preserve the fabric of the major French cathedrals and churches. We now know that Viollet-le-Duc went far beyond his original preservation brief and expanded his influence to what might now be termed as restoration and subjective intervention to establish an idealised completion of church buildings that had never previously existed. The creation of a complete former state was for him a prime concern and consequently, his work was criticised as "the vandalism of completion". He saw the role of the restorer as one of "re-establishing a state of completion which may never have existed at any given moment in the past" (Earl, 1997).

In England the appearance of the cathedrals of Durham, Hereford, Lichfield, St Alban's and Salisbury, as well as Westminster Abbey, were also 'improved' to conform to an idealised vision of what Gothic architecture should demonstrate. In so doing, the developed style achieved over many centuries was superseded in an effort to subjectively correct the design but with no recognition of the importance of the historic record accrued in that building.

Like Wyatt, the architect Sir George Gilbert Scott (1811–1878) was also heavily criticised for his work on a number of cathedrals, including St Alban's Abbey. Here he was project architect from 1860 onwards and undertook very major works including restoration of the medieval floor and substantial alterations to the walls and roof. However, Scott denied his work was anything other than preservation and considered himself to be a "conservative restorer" (Pickard, 2000). Pevsner (1976) suggests that Scott did much to enrich the fitments of the cathedrals which he worked on and at a time when many medieval buildings were in a "state of shocking neglect". However, Scott admitted in his *On the Conservation of Architectural Monuments and Remains* of 1864 that "we are all offenders in this matter" (cited in Pickard, 2000).

Earl (1997) also points out that people like Viollet-le-Duc were often averting total loss and were instrumental in creating the preserved townscape settings now seen as the norm.

By the mid-nineteenth century two philosophical debates had arisen as to whether buildings should be either preserved or restored. According to Earl (1997) restoration refers to the creation of the most desirable or perfect form of a building. Although it can refer to repair or general improvement, it may also mean the process of bringing the building back into a former state, for example, the 're-medievalising' of a building. Preservation on the other hand, means preserving intact what has been inherited from the past by rejecting all unnecessary interventions.

This philosophical debate inspired William Morris to set up the Society for the Protection of Ancient Buildings in 1877 in an attempt to encourage conservative repair. SPAB believed that restoration would falsify the history of a building and that the architecture of the past should be preserved with the minimum of alteration so that the forms and techniques could be studied. These basic principles were encompassed in the SPAB manifesto of 1877 and the basic tenets of conservation philosophy founded in this are summarised by Earl (1997):

1   We are merely custodians of historic buildings and should treat them accordingly.
2   Effective and honest repair is a prime consideration.
3   There should be minimum intervention.
4   New work should be clearly identifiable and should not imitate a past style.

SPAB established a new approach to building repair and conservation. This had wide appeal as Pickard (2000: 146) states:

> The wider influence of the Society certainly assisted the process of changing attitudes away from restoration in favour of conservative repair on an international level. The manifesto was a first attempt to set down basic ideals which guided other bodies in developing the philosophy of conservation.

This therefore formed the starting point for the conservation movement. William Morris very pointedly stated:

> These buildings do not belong to us only … they have belonged to our forefathers and they will belong to our descendants unless we play them false. They are not … our property, to do as we like with. We are only trustees for those that come after us.

This idea of stewardship has also become central to conservation philosophy.

Chapter **42**

The twentieth century and establishment of conservation charters

The development of international charters has proved pivotal in the active conservation of historic buildings. Building on the foundations of SPAB, these charters outline the correct approach to conservation to be undertaken.

ICOMOS

The International Council on Monuments and Sites (ICOMOS) is a non-governmental organisation which has worldwide membership. Focused on the cultural heritage, they also act as special adviser to UNESCO on World Heritage sites of which there are currently 26 in the United Kingdom.

The Venice Charter was published by ICOMOS in 1964 and is concerned with the "basics of modern conservation". The articles set out in this Charter consider conservation, restoration, excavations and the publication of reports. As an international charter, a framework is set out but each country has a duty to apply it as appropriate given its particular beliefs, cultures and traditions.

The Venice Charter was the first one to be established by ICOMOS. A number followed, including the Florence Charter on historic gardens (1982) and in 1987 the Washington Charter addressed the issue of historic towns and urban areas. However, one of the most important charters was the Burra Charter (1979, current version 1999) which is an Australian charter concerned with cultural significance. A definition of cultural significance will be examined later in this part.

Chapter **43**

Legislation

In the United Kingdom, protection of historic fabric is achieved through the listing of buildings, scheduling of ancient monuments and the designation of conservation areas. However, the legislation varies within the United Kingdom and will obviously be different across Europe and beyond. For any practitioner in conservation, it is vital to understand the mechanisms which offer protection to historic structures in order that appropriate consents are applied for when undertaking any work. As Earl (1997) states in respect to the criteria necessary for effective control, legislation must incorporate the following requirements:

- a definition of what is to be preserved – a list;
- a method by which authorities might be alerted to possible danger – a requirement for notice;
- a way of permitting harmless or desirable works – a consent process;
- effective sanctions against offenders – a punishment process.

Legislation changes continuously and is extensive. This book merely gives an overview of the legislative powers and the reader is encouraged to read wider on the subject and to seek relevant local legislation. The historical context is important to understand because the development of the legislation has a bearing on the historic buildings that we now have.

During the nineteenth century the conservation movement was born through the work of William Morris and SPAB. Following these efforts, the *Ancient Monuments Protection Act 1882* became the first legislation aimed at protecting historically important buildings. However, only 68 monuments were "scheduled" and there were no compulsory purchase provisions. Despite the obvious limitations of this initial legislation it signified a change in attitudes to historic buildings and paved the way for significant change in the twentieth century.

The twentieth century opened with the revised *Ancient Monuments Protection Act 1900* which gave protection to medieval, as well as prehistoric, buildings and in 1908 the three *Royal Commissions on the Historical Monuments* (of England, Scotland and Wales) were established. Earlier legislation was improved through the *Ancient Monuments Consolidation and Amendment Act 1913* which both listed monuments of importance and also granted them protection as they could be acquired and maintained at public expense (Boulting, 1976). However, it was 1947 before the most comprehensive legislation the *Town & Country Planning Act* was introduced. This made the compilation of lists of buildings of architectural or historic interest a statutory duty. A national survey of all historic buildings was initiated using investigators who inspected towns and villages across the United Kingdom to identify the buildings of greatest importance which needed to be listed. The first list, completed in 1968, was the first time the exercise had been carried out so there were naturally many serious gaps in the survey which can be attributed largely to the lack of information available to the investigators.

The principles of listing

The legislation for statutory listing is different in Scotland from England. In England, under the Planning (Listed Buildings and Conservation Areas) Act 1990, buildings which are considered to be of "special architectural or historic interest" are placed on a statutory list. It is then a legal requirement to obtain listed building consent prior to carrying out any alterations to a listed building.

The current listing system in England defines the following categories of listing importance:

- Scheduled Ancient Monument
- Grade I
- Grade II
- Grade II*.

Buildings may be included in the list for a number of reasons. They may be of particular architectural interest, demonstrating a building type or construction technique. They may also be of importance because of their association with a particular place, event or person, for example, Robert Burns' cottage in Ayrshire or as the work of an important artist or architect, such as Charles Rennie Mackintosh. Inclusion may also be related to a particular aspect of history, such as cotton mills at New Lanark. Buildings may also demonstrate group value, such as the terraces found in Bath and Edinburgh. The age of the building is also a consideration and the older, intact examples are likely to be listed. More recent buildings are less likely to be listed and listing of any building less than 30 years old is rare unless the building is a particularly special example or is under threat.

However, the system in England has been under review and a White Paper, *Heritage Protection for the 21st Century* (Department of Culture, Media and Sport, 2007) has been published following consideration of the system and its effectiveness. The publication outlines the aims of legislation in protecting the heritage of England stating: "An effective heritage protection system needs to strike a balance between protecting what is important and enabling appropriate change…. If we seek to prevent all change, the heritage protection system quickly becomes devalued and unworkable."

Under the proposals for review the listings would have changed in England and Wales to Scheduled Ancient Monument, Grade I and Grade II, combining Grade II* with Grade I. However, the DCMS concluded that the existing system is well understood and the proposals for change will not now occur. Listing grades will therefore remain unaltered but will be referred to as G.I, G.II and G.II*. Scheduling of National Monuments will also remain as existing but will be designated Grade I; all grading will be subject to English Heritage review. The principles put forward in the DCMS White Paper also recommend that all listing and scheduling be placed under the guidance of English Heritage, including both land-based and maritime archaeology, together with battlefields, parks and gardens. World Heritage sites remain under the aegis of UNESCO. The proposals also make recommendations for heritage management plans to be implemented for larger and important sites. This is discussed in Part 7.

In Scotland, under section 1 of the Town and Country Planning (Listed Buildings and Conservation Areas) (Scotland) Act 1997, a statutory list of buildings of special architectural or historic interest is compiled by Historic Scotland for the Scottish Ministers. These buildings are afforded statutory protection, making it an offence to alter them without permission. Listed building consent is required under sections 6 and 7 for the demolition of a listed building or for any alteration or extension which would affect its character as a building of special architectural or historic interest. This also covers objects or structures fixed to a listed building or within its curtilage which existed before 1 July 1948. The legislation is not restricted to buildings, but includes non-habitable structures such as bridges, ruins and monuments.

The Memorandum of Guidance (Historic Scotland, 1998b) states that the buildings to be listed include all those built before 1840 which are largely original in form. More recent buildings may be listed if they have certain characteristics worthy of preservation. Buildings of particular types or which display technological advancement or which are associated with famous people should also be considered.

There are three categories of listed building in Scotland as defined in the *Memorandum of Guidance*: Category A, buildings of national importance; Category B, buildings of regional importance; and Category C(S), buildings of local importance. This latter category replaces the old non-statutory Category C. In addition to individual listings, group listings are used to stress the group importance of buildings which relate together, such as in townscapes. This grouping does not alter any individual listing category but emphasises the association of that building with others in the group.

In England, Scotland and Wales there are 470,000 listed buildings in total. England has 370,000 listed buildings, of which 92 per cent are Grade II. In Scotland, there are 47,000 listed buildings and the majority are in Category B at 60 per cent, with 32 per cent in C(S) and only 8 per cent in Category A. Wales has over 25,000 listed, of which over 90 per cent are Grade II.

Conservation areas

The 1947 Town and Country Planning Acts heralded change, but it was not necessarily a rapid change. However, during the 1950s and 1960s further support for historic buildings was galvanised through the Civic Trust movement. The Civic Amenities Act 1967 gave local authorities a statutory duty to designate conservation areas. A national re-survey of listed buildings was also initiated in 1969 by the Ministry of Housing and Local Government.

Under legislation, local planning authorities in Scotland and England must determine areas which they deem to be of "special architectural or historic interest, the character or appearance of which it is desirable to preserve or enhance". The exception is that within London, English Heritage may designate conservation areas in consultation with the local councils there.

Conservation areas are extremely varied but may include groups of buildings, streets, town walls, rivers and open spaces and locations of historic importance.

Listed building or conservation area consent is not required for the alteration or extension of unlisted buildings in a conservation area, but consent is required for demolition of unlisted buildings. Planning authorities may, however, extend their powers within a conservation area by restricting minor works under an Article 4 Direction which limits otherwise permitted development rights in order to prevent the erosion of the character of the area. This includes extensions such as porches and dormer windows as well as the alteration of doors and windows. The Historic Scotland *Memorandum of Guidance* (1998b:

53) highlights the importance of maintaining standards, particularly for unlisted properties:

> Unlisted buildings, although often humble, have their own important part to play within a conservation area, frequently forming a sympathetic and complementary backdrop for their more pretentious listed counterparts. Any erosion in standards permitted for work to unlisted properties will inevitably in the long term seriously affect the architectural quality of the conservation area of which the buildings form a constituent part and should consequently be actively discouraged.

Trees are also recognised as making a significant contribution to conservation areas. Six weeks' notice must therefore be given to the local authority if it is proposed to cut down or lop a tree within a conservation area. This is to allow the authority time to assess the tree to decide if a Tree Preservation Order is required. Such an Order will apply to trees which are of significance within a district in order to ensure that they are protected from removal or damage.

It is evident that conservation areas aim to maintain standards of townscapes by protecting buildings, spaces and vegetation which all make a contribution to the quality of the place. However, in addition to designating a conservation area, local authorities also have a duty to introduce proposals which will enhance and protect conservation areas. Positive management is essential to raise the quality of the built environment. With in excess of 8,000 conservation areas in England and over 600 in Scotland, this is a major responsibility for local authorities throughout the country.

Ecclesiastical exemption

Churches in current and continual use for worship are not subject to the same statutory controls as other listed buildings because they have ecclesiastical exemption. Despite this exemption, permission is still required for the demolition of churches or for demolition of churches in conservation areas. This is because they will lose their designation as places of worship prior to demolition and therefore do not fall within the remit of ecclesiastical exemption (Kent, 1999). In Scotland and England there are over 40,000 listed churches, with an additional 3,500 non-conformist churches (Kent, 1999).

In Scotland, a three-year pilot scheme operated to investigate the effect of applying listed-building control to the exterior of churches. This was a partnership between Historic Scotland, local planning authorities and the Scottish Churches Committee. Since the end of the pilot scheme in 2004 a voluntary code has continued to apply to churches which would require listed building consent if ecclesiastical exemption were not in place. The aim is to maintain the heritage of Scotland's church buildings.

In England a similar internal scheme operates with a code of practice whereby congregations must notify a relevant and independent decision-making body of their intended alterations. This body has expertise in church buildings but will also consult with relevant parties such as English Heritage and local planning authorities. Usually these independent bodies are sub-groups or committees within the overall religious organisation. Their powers are subject to regular scrutiny and re-evaluation periodically by central government to ensure effective authority and control.

Ecclesiastical exemption continues to be under review and therefore approaches may alter in the future.

Historic parks and gardens

In addition to the protection of buildings and monuments, there is also protection offered to historic parks and gardens. English Heritage maintains a register of historic parks and gardens and use a nine-point criteria to determine whether a garden should be included. Although this does not offer the statutory protection that listed buildings have, where a planning application is to affect a registered property the local planning authority must take into account its status. In Scotland there is an Inventory of Gardens and Designed Landscapes which is maintained by Historic Scotland.

Important landscapes include gardens such as those which are part of large country houses or other buildings such as hospitals and also cemeteries.

Government guidance

In addition to legislation, there are various types of government guidance which aim to supplement the planning legislation and ensure that there is a consistent approach by the appropriate authorities. The three main types of government guidance are:

- **Circulars**: these provide updated information on procedural and legislative matters. For example, Circular 1/1998 provides updated information on changes to the Use Classes Order.
- **Technical guidance – planning policy guidelines (PPGs)/**

national planning policy guidelines (NPPGs): national planning guidelines (NPGs) were first established in Scotland in 1974 to develop guidelines for the North Sea oil and gas industry. Like their counterpart in England and Wales, known as planning policy guidelines, these are not location specific but contain broad policies relating to national land use and development policies. The advice contained in these guidelines is closely linked with the development plan system. Cullingworth and Nadin (1997) state that PPGs impact on planning practice because they "are important material considerations in development control and have a determining influence on the content of development plans". PPG 15 is the most important document relating to the historic environment, the Scottish equivalent being NPPG 18: Planning and the Heritage.

In Scotland NPPGs are now being replaced by Scottish planning policy documents (SPPs). These will run alongside existing NPPGs until they are all replaced. SPP1: The Planning System replaces NPPG 1 and gives an overview of the planning system.

- **Planning advice notes**: these cover particular issues in planning and give advice on good practice. They cover all aspects of a particular type of planning issue, such as small towns, retailing or telecommunications.

Chapter **44**

The role of amenity societies

In addition to legislation and guidance, there are a number of groups who provide an invaluable service in protecting the historic environment. These groups are varied in their focus but they perform vital functions in a number of areas. First, they raise awareness about the importance of particular buildings. Second, some groups offer training and advice about particular aspects of historic buildings. Third, they may be involved in campaigns to save threatened buildings which they consider to be of importance. Finally, certain groups may be better able to access funds in order to carry out conservation work.

The following is a summary of these groups:

Architectural societies: these include groups who have a particular interest in a specialised building type. This group includes the Victorian Society, the Twentieth Century Society, the Georgian Group, the Ancient Monuments Society, the Friends of Friendless Churches, the Architectural Heritage Society of Scotland and the Society for the Protection of Ancient Buildings. They are special-interest societies which focus their efforts on particular building types and can often provide specialist advice and encourage research and education through their publications and newsletters.

Building preservation trusts: these groups are registered charities and may either be revolving or single trusts where they deal either with a succession of building projects or with a single, one-off project. They can access funds either through the Architectural Heritage Fund or through the Heritage Lottery Fund. Projects often involve a building which is at risk of being demolished and the action of such a group can save and restore a building and give it a viable re-use. Examples include the Scottish Lime Centre, Charlestown in Fife and Cromford Mill in Derbyshire. One of the most well-known building preservation trusts is the Landmark Trust who care for many highly unusual buildings and lease them out as holiday homes, so helping to pay for their up-keep. Their properties include the Charles Rennie Mackintosh building in Comrie, Perthshire and The Pineapple in Falkirk.

National Trust / National Trust for Scotland: the National Trust was founded in 1895 and cares for 300 historic houses, 49 industrial mills and numerous nature reserves, woodlands, farms and villages. Their membership and entry fees help to pay for the up-keep of their substantial portfolio of properties. They also encourage conservation and education.

The Civic Trust / Scottish Civic Trust: the Civic Trust was founded in 1957 (1967 in Scotland) by Duncan Sandys in response to the destruction and deterioration of many towns. The Civic Trust Movement was instrumental in bringing the Civic Amenities Act of 1967 to fruition, which then established Conservation Areas. The Trust is a network of individual societies who are concerned with their local area but with the support of a national organisation. They carry out education and campaigns of improvement to urban areas. They also help to protect threatened buildings and to maintain standards in town centres.

Chapter **45**

Conclusions

Current legislation and accepted conservation doctrines have their origins in the philosophical debates of the nineteenth century. William Morris and the Society for the Protection of Ancient Buildings took the debate forward by establishing a set of principles which still applies over 100 years later. These principles have been given international recognition through the adoption of various charters such as the Venice and Burra Charters.

The planning system aims to control development without hindering economic progress. Building owners must apply for planning permission before proceeding with development works, and in conservation areas and with listed buildings the controls are, rightly, stricter.

Professionals dealing with historic buildings must be conversant with the latest planning legislation to ensure that the correct permissions are received prior to undertaking any work.

Part 7
Conservation in practice

Part 7

Introduction

The previous parts have examined architectural history, materials and legislation. An understanding of these inter-related subjects is the foundation for conservation. The context and construction of buildings, together with the legislation that aims to protect historic buildings forms the core of understanding for any conservation practitioner. This part develops these themes and considers conservation philosophy and its practical applications through the use of conservation plans and the consideration of particular and sometimes challenging issues with historic buildings, such as dealing with change of use for redundant buildings and caring for ruins.

In 1965 ICOMOS was founded as a non-governmental organisation whose remit was to provide a forum for all those involved in the conservation of cultural heritage. It is the internationalisation of conservation response that has made for a truly worldwide, co-ordinated approach to conservation. Its formation was inspired by the 1964 Athens Charter and remains the ideal forum for the formulation and promulgation of a consensus on international conservation response, its aim being to improve awareness of "the theory, methodology and technology applied to the conservation, protection and promotion of the worth of monuments and historic areas". ICOMOS acts as a forum for exchange of ideas across international professional boundaries. It has forged links with other international organisations such as the Council of Europe (formed in 1949) and UNESCO.

The Council of Europe defines its remit as:

> to foster the advancement of heritage protection and enhance policies within the framework of a pan-European project of cultural and social development, and to develop a model for European society where the right to a heritage, that is, the right to a memory and to better living environment, could

constitute a new generation of human rights, after political rights, social rights and the right to information.

The United Nations Educational, Scientific and Cultural Organization (UNESCO) seeks to encourage the identification, protection and preservation of cultural and natural heritage around the world considered to be of outstanding value to humanity.

These three organisations encapsulate the international response to conservation promotion and seek, jointly, to further the cause of international heritage and are fundamental to the promulgation of conservation philosophy and the practical application of it. All three organisations adopt the fundamental principles of conservation that are discussed below.

Chapter **46**

Principles, ethics and criteria of conservation

The basic approaches to conservation were established through the SPAB manifesto of 1877. The ethics or criteria of conservation are as follows:

- minimal intervention
- minimal loss of fabric
- minimal loss of authenticity
- absence of deception
- reversibility.

These fundamental principles or criteria are discussed by Bell (1997) in Historic Scotland Technical Advice Note 8: *The Historic Scotland Guide to International Conservation Charters*, and are accepted, via these charters, as the core international principles of conservation response. Selected quotations from the various international charters are used to expand the definitions given below.

Minimal intervention

> Conservation is based on a respect for the existing fabric and should involve the least possible physical intervention.
>
> Burra Charter, 1979

This means that the least amount of intervention should be undertaken when carrying out work to a historic building. It may be tempting to carry out additional works to 'modernise' or upgrade a property. However, only works which are strictly necessary should be implemented. The principle of minimal intervention is the by-word of conservation. By complying with this principle the conservator will avoid the danger of unnecessary destruction of fabric and loss of authenticity that would otherwise diminish a historic building.

Minimal loss of fabric

> Conservation should show the greatest respect for, and involve the least possible loss of, material of cultural value.
>
> New Zealand Charter, 1992 (see also Deschambault Declaration, 1982)

This is synonymous with minimal intervention. In this case, the maximum amount of historic material possible should be retained. It may be necessary to replace some worn out or damaged materials, but sound materials should always be retained. For example, where a traditional sash and case window exists, small repairs to rotten window-sills can be undertaken, but the entire window should not be removed and replaced because of what is essentially a minor repair.

Minimal loss of authenticity

> Imbued with a message from the past, the historic monuments of generations of people remain to the present day as living witnesses of their age old traditions.... The common responsibility to safeguard them for future generations is recognised. It is our duty to hand them on in the full richness of their authenticity.
>
> Venice Charter, 1964

Authenticity is a difficult term to accurately define as it is often confused with originality. An authentic element may not be original but is 'authentic' of its period of introduction. Original is an element that was incorporated at the time of construction of an asset. So, a window frame introduced in 1850 to a building dating to 1700 is not original to the building when it was first created but will be authentic of its period of introduction. As Bell (1997) defines by reference to the Declaration of Dresden: "All original fabric is authentic but not all authentic material is original."

However, identifying what is original is not easy. Tiedsell et al. (1996: 173) suggest that the ability to establish what is authentic will vary from building to building and state that "Authentic and historic fabric may be obvious and untainted by years of intervention. However, more often it will be the case that there have been repairs, alterations or additions and removals." The conservation practitioner therefore needs to investigate a building carefully to identify these alterations.

Absence of deception, or legibility of intervention

This was a philosophy central to the values of John Ruskin and Augustus Pugin. They rejected architectural deceits where materials were made to appear different from what they actually were, such as marbling of timber to make it look like stone. This also extends to not making new work look as though it is old as this might be considered a dishonest repair. The guiding principle here being that intervention work should be obvious to the trained eye without detracting from the overall impression of completeness to a lay observer. As the Venice Charter defines: "Replacement of missing parts ... must be distinguishable from the original so that restoration does not falsify the artistic or historic evidence."

For example, if a brick is replaced within a historic wall, the new brick should not be distressed in order to make it appear as though it is old, but instead it should be a suitable match in terms of colour, texture and size so that it is honest without, necessarily, being obvious to the lay, or casual observer.

Reversibility

> The use of reversible processes is always to be preferred to allow the widest options for future development or the correction of unforeseen problems, or where the integrity of the resource could be affected.
>
> Venice Charter, 1964

This is also important to consider, particularly when introducing new elements to a building. The Appleton Charter of 1983 states that "[t]he use of reversible process is always to be preferred to allow the widest options for future development or the correction of unforeseen problems, or where the integrity of the resource could be affected".

Alterations must therefore be incorporated with the capacity to be removed with minimal damage, loss or destruction of fabric or structure. Circumstances may change in the future for a building, such as a change of use or development of a new technology that might facilitate a more appropriate alternative, and this potential must be considered at the outset. For example, a building which is open to the public may require, for example, modifications in order to comply with the Disability Discrimination Act 1995. However, inclusion of ramps, lifts and handrails can be visually and structurally intrusive and these may require creative solutions such that they can be installed safely, perform their purpose and then be removed at a later date without damaging the building. This can be particularly challenging when proposing the inclusion of modern services such as heating, lighting and plumbing in historic buildings.

Conservation is about differentiating between *appropriate* as opposed to *inappropriate* intervention. As Humberstone (1997) states: "Conservation therefore relies on the moral underpinning of a set of principles which are sustained by reason and argument, and which distinguish the appropriate from the inappropriate." It is about facilitating and managing change without loss of authenticity, record and value. In this context it is important to be sure of the possible implications that 'restoration' may have as opposed to conservation.

In Britain there is a presumption against restoration. BS 7913: *The Principles of Conservation of Historic Buildings* (1998) defines the particular terms and definitions used to define conservation. It defines *restoration* as:

> Alteration of a building, or part of a building or artefact which has decayed, been lost or damaged or is thought to have been inappropriately repaired or altered in the past, the objective of which is to make it conform again to its original design or appearance at a previous date.

Paragraphs 7.3.2.1–7.3.2.3 define the case for and against restoration, stating that there should always be a presumption against restoration which involves returning an asset to a former state where that involves any degree of subjective interpretation. Restoration might be appropriate if there is incontrovertible documentary evidence of such a former state or, "[where] buildings of formal, or classical, design" show detailed evidence of where there is a void or lacuna in an otherwise complete design or, "where the existence of a known or proven design for the missing building, or element, feature or detail" can be incontrovertibly identified. The important process here is the ability to establish, beyond reasonable doubt, the existence of a former state, condition or design.

It is this philosophy that defines the British approach to a basic principle of conservation – that of only working within a structure that is based on detailed research, analysis and total comprehension before undertaking any work involving anything that might damage or diminish the built heritage. It must be recognised, however, that this approach is a particularly Western-world philosophy which may differ in other parts of the world. Japan, for example, may periodically dismantle historic buildings to check for deterioration and rectify or repair any developing defects in order to maintain an ideal or pristine vision of an asset.

Although 'restoration' is questioned by British conservation philosophy, a different approach may be made by other national philosophies. In Russia, for example, at St Petersburg, the magnificent Catherine's Palace, destroyed during the Russian campaign of the Second World War, has been lovingly restored as a matter of national pride. Warsaw, destroyed during the same war, has been restored to its former glory. Both these restoration projects relied on detailed research and comprehensive records of the former but destroyed state, thereby permitting an academic restoration exercise to take place.

Chapter **47**

Conservation plans and cultural significance

When dealing with a historic building it is vital that an appropriate plan is put in place so that the conservation philosophy set out previously is central to the approach adopted. This can be achieved through the use of a conservation plan.

A conservation plan is essentially a tool to allow a historic site to be managed effectively. In doing so, it will consider the cultural significance of that site and have built into it the need to protect this significance. A plan will also identify particular elements that may be especially vulnerable to change so that suitable measures are put in place to minimise any impact caused by change or intervention.

A conservation plan will be undertaken prior to any work being carried out to a building such as an extension or change of use, but it can also form a more fundamental part of the management of a building by informing day-to-day maintenance regimes. They may also be a requirement of a grant application, such as for funding from the Heritage Lottery Fund.

Conservation statements are basically early statements of intent which are prepared at the beginning of any intervention process and are the first attempt at planning conservation work, repair or maintenance. They are a short-hand version of what will, eventually, become conservation plans. The statement will identify cultural significance and condition and may also focus on areas where more detailed study is required. They might also identify early estimations of cost for intervention work and where funding might be sought.

Conservation plans follow on directly from the conservation statement and will build on the knowledge gained from the preparation of the previous statement. Their preparation will involve considerable research in order to fully understand the significance of an asset. Conservation plans must also address future management policy and how the impact of regulatory controls may affect the assets and its significance. An example of this might be inclusion of disabled access with only minimal impact and loss of significance. This is a particularly important function of the conservation plan and should address all issues relating to the on-going management, administration and use of an asset.

Part of the conservation plan process is to define the future management and decision-making structure and to identify particular management responsibilities. These responsibilities may lie with an individual but are also part of the whole management process. These important management decisions must be incorporated into any business plan for the asset so that maintenance of significance is the primary driver in any economic and administrative decision. The conservation plan should be adopted as a management document and should inform any future decisions. Plans should also be used to formulate budget expenditure and financial allocation, identifying a priority process of repair and maintenance balanced against income and funding sourcing.

The Heritage Lottery Fund in *Conservation Plans for Historic Places* (1998) suggests the contents of a conservation plan should include the following elements:

1   contents page;
2   summary of the asset and its heritage merit together with how it is vulnerable and appropriate policies to be adopted;
3   background to the asset and reasons why a conservation plan is needed;
4   understanding of the site with detailed descriptions, maps and images;
5   assessment of significance, detailing what aspects of the site have heritage merit;
6   definition of vulnerability including condition, use, location and site history;
7   conservation policies with a vision for the site;
8   implementation and review procedures.

Assessing cultural significance is central to all conservation plans. The Burra Charter established the principles of cultural significance. Article 2 of the Charter states: "The aim of conservation is to retain or recover the cultural significance of a place and must include provision for its security, its maintenance and its future."

Here, "cultural significance" is defined as "aesthetic, historic, scientific or social value for past, present or future generations".

These various elements allow individuals to consider what may be important to them, the cultural significance of *their* place. This will vary from community to community depending on that community's own beliefs and values. These values may be present in differing levels of importance, but are a synthesis of all in most assessments of cultural significance. It is therefore vital, prior to planning any works of intervention involving the historic environment, to identify and carry out a detailed analysis of the various elements that contribute to an asset's significance in cultural terms.

How and why a place is important must be the first question addressed before any works of intervention are undertaken. Part of the process will involve public consultation. Local involvement, asking people's opinions on the asset and its worth is an essential element in all conservation intervention planning. This not only involves the public in the process but also improves their knowledge and understanding of the asset and its cultural worth. This forms a symbiotic process of assessing, evaluating and understanding which is a vital part of conservation intervention.

An example of where a conservation plan was used is at Newhailes House near Edinburgh, which is now owned by the National Trust for Scotland. The house was designed in 1686 by architect James Smith (1645–1731) in a Palladian style and "is important as a precursor to the kind of Palladian villa that was later popular throughout Britain, Ireland and North America" (National Trust for Scotland, 2004). In the early eighteenth century the house was extended, including the addition of a library and the setting of a designed landscape around the house. Features in the house reflect the rococo period with use of shells, including a shell grotto in the gardens.

The National Trust for Scotland acquired the house in 1997 and they state (2004) that they

> undertook historical research and survey work into its origins, development and significance, the better to inform our conservation plan. The specialist survey work included such diverse elements as landfill and topographic reports, glazing and paintwork analysis, and surveys on collections such as the clocks and ceramics.

This plan then allowed appropriate decisions to be made in order to protect and ensure the future of this nationally important house. Holder (2002) summarised the approach adopted by the National Trust for Scotland, saying that they identified that the most significant element of the house was its "mellowness" and this meant that they sought to do "as much as necessary but as little as possible". Therefore, the intervention work that was undertaken involved only those alterations that were required for specific purposes. This included those such as health and safety or for staff and visitor accommodation. This approach is exemplified in the minimal alteration to the external handrail.

47.1   Newhailes House, Edinburgh

Conservation heating needed to be installed in order to control heat and humidity and a fire suppression system was installed because of the particular value of the house and its contents. These had to be installed in floor voids, but this was undertaken in a fully informed way following the conservation plan and only floorboards which had been previously lifted were allowed to be removed. Despite the adoption of a philosophy of minimal intervention, works, including structural, were required to ensure the long-term future of the building although, again, these were fully informed by the detailed investigation that had been undertaken.

In allowing access to the building for the public, the visitors need to be carefully managed. This involves allowing only certain numbers through, escorted by a guide and ensuring that the minimum impact on the building by visitors occurs. Again, the conservation plan has ensured that the site is effectively managed.

47.2   Minimal handrail repair to the main entrance and access

Conservation management plans are likely to be more applicable to large sites where there are buildings in continued use and which may be constantly changing and developing. Two examples of such large sites but with differing history might be Holkham Hall in Norfolk and the University of East Anglia (UEA) in Norwich.

Holkham Hall is an eighteenth-century Palladian country house and managed estate. The house is a major tourist attraction and its contiguous estate is currently managed as farmland and a country-pursuits park. Associated properties within the estate include housing for estate workers and functioning ancient and modern agricultural buildings as well as changing but relatively constant land uses. Its potential for long-term change is likely to be relatively limited and its pattern of use will probably remain largely unaltered. The adoption of conservation management will present as less of a complex issue than will, say, a site in constant change and development such as a university campus like that at the University of East Anglia.

The UEA was created as a new university in 1962 under the design of Denys Lasdun. Admitting its first student in 1963 it was given Grade II* listed status in respect of Lasdun's "ziggurat" student accommodation, its teaching wall, reference library and main square. It is a constantly developing site reacting to and accommodating new subjects, additional students and constantly expanding accommodation needs. It is on a limited site with restricted ability to expand over the protected landscape of the campus across the Wensum valley.

47.3   Ziggurat student accommodation, UEA, Norwich

These two disparate but nationally significant sites were chosen as "test beds" for the government-thinking on the future of heritage under the current planning system specifically because of the differing patterns of use. This then allowed an analysis of those differing uses and consideration of how a listed-building control system might similarly be changed to accommodate the necessary control and monitoring of use.

Large sites, in the future, will have to agree their forward management plans with central government so that the heritage might be best protected but permitted to develop within a known and approved development plan over a period of currency. This, in simple terms, is the method and manor of conservation management plans as proposed under the government review, *Listing is Changing*, published in 2005.

Chapter **48**

Investigating and understanding buildings

If a conservation plan is to be undertaken, a building must be fully investigated as part of that process. This will involve a series of steps in order to arrive at an informed decision and this must be carried out in a systematic and analytical way.

Such a structured approach to pre-works investigation is essential and is outlined by Bell (1997). This approach will involve:

- recording the building as found;
- investigation of the site's significance and condition, including study of the building as a primary source and documentary records as a secondary source;
- the public must also be consulted to establish their view and understanding of the asset;
- these actions will lead to the preparation of a Statement of Cultural Significance; from which
- a plan of action may be prepared covering short- and long-term plans for maintenance or change (the conservation plan or conservation management plan);
- a detailed record of all evidence found and all interventions undertaken will need to be made and placed in public archives for future reference.

Record

Investigate

Consult

Prepare statement of cultural significance

Prepare plan of action

Implement work

Record

48.1

Recording and background research

One of the first and most important steps in understanding a building is to investigate its history and development. This will be a two-step process involving a detailed examination of the building together with background research using secondary sources.

A building is the primary source of information, together with any associated documentation such as within its own library or archive. The building as a source of information may be complex, as in the case of very old and frequently adapted structures, or may be incomplete, as in the case of a ruin or archaeological site.

A building survey will be necessary to provide detailed information about the construction. While many practitioners will be sufficiently skilled to undertake such work, there may be particular investigations which will require specialist advice. It is not possible to be an expert in every field and indeed, a professional will recognise the limitations of their own knowledge and employ a suitably qualified expert to carry out the necessary work.

The types of work undertaken may include the following:

- **Measured and levelled survey**: from which detailed drawings of the existing building and fabric may be prepared.
- **Archaeological survey**: this may be of above or below ground structures and, in some cases an archaeological investigation may be a stipulation of development.
- **Photographic recording**: this may include specialist photography such as rectified photography or photogrammetry. These can be particularly useful in identifying areas of previous work and alterations and so can assist in understanding the development of complex buildings.
- **Dendrochronology**: this uses tree-ring growths to establish the date of timbers used in buildings.
- **Paint analysis**: paint scrapes are taken to identify layers of paint and their colours and compositions by examining the samples under a microscope.
- **Infra-red survey**: this uses specialist equipment to investigate the nature of hidden structures.
- **Visual investigation**: this can use specialist equipment such as boroscopes and miniature cameras.

It will be rare that a building will give up all the answers. It is therefore usually necessary to carry out further detailed research using various documentary sources. These may be found in a variety of locations including local and national libraries, museums, national and local archives and, increasingly, the Internet. Useful records may vary from maps and photographs to architects plans or museum artefacts. These will help to build-up a picture of the history of the building and the sequences of development that have occurred over time. Such records can help to clarify issues highlighted by an examination of the building fabric. For example, historic photographs may indicate architectural details which have since been lost such as finials or crestings and historic maps may be used to show when a building was first located on a particular site.

Possible sources for information are as follows:

- local newspaper offices and archives
- church records
- local and national museums
- listed building records and local authority archives
- archaeological studies and reports
- national and local records offices
- web-based information
- Site and Monuments Records and Royal Commission on Ancient and Historic Monuments Scotland (RCAHMS)
- published works, such as texts on the history of a local area, or contemporary accounts
- research reports and theses
- national body records such as Historic Scotland, English Heritage, National Trust, CADW or EHSDE in Northern Ireland
- photographs, prints and works of art – both locally and nationally stored
- Ordnance Survey and historical maps
- geographical, geological surveys and reports
- historical manuscripts and records, including, possibly, historical account sheets associated with the subject building
- local libraries
- local amenity societies and historical societies who may hold their own records or undertake research on particular aspects of the community and its environment.

The above is not an exhaustive list but does provide a useful indication of the vast array of potential sources of information available. The whole process of investigative detective work can, in itself, be both rewarding and interesting and lead to a stimulation of enthusiasm for the subject in its own right. However, such research can also be time-consuming, particularly if the site is very complex, although the increasing availability of material on the Internet makes identifying and accessing documents considerably easier than in the past. The digitisation of records means that many of these documents can be viewed remotely over the Internet and this in turn preserves fragile documents that would otherwise require handling.

Consult the general public and other interest groups

A fundamental requirement in the preparation of conservation plans is that of consultation with stakeholders and the general public. This will canvass views and encourage public involvement in the planning and implementation of conservation work. This action has the added advantage of raising awareness of an asset and encouraging uptake of greater knowledge and involvement by those who might otherwise feel excluded from commitment to, and involvement with, their local heritage.

The concern and interest that the general public has for their historic buildings should not be underestimated. People feel strongly about their local area, its history and the value of their communities. It is therefore important that the public is consulted where decisions are being made about a building of importance to a community. This may be particularly important where a building engenders strong feelings, such as a church, school, hall or other community building. It is vital that the various opinions are sought and accommodated, as far as possible, in any decision-making process.

Other bodies may have a more focused interest such as societies specialising in particular buildings. For example, the Twentieth Century Society is, clearly, concerned with the conservation of modern buildings. Others may have a wider range of interest, such as the Civic Trust.

Although the public may be aware of the importance of a building, they may not know the exact reasons why such a building is significant. It is therefore the duty of the conservation professional to ensure that the appropriate education, promotion and publicity is carried out. Improving the knowledge and understanding of the public will add to their enjoyment of it and by engendering support through understanding, the public can become the best advocates of historic buildings.

The educational value of the heritage is well recognised and involvement with local schools is to be encouraged within the process of consultation. This will generate a symbiotic response via education and enlightenment of future generations whose commitment to the heritage will be necessary in order to foster its future use and evaluation by society.

Implementation of works

Once it has been established that works are required either in maintenance, repair or in more major work such as a change of use or conversion, then the process of implementation will require the procurement of building work along with the choice of an appropriate contractor and contractual arrangements.

It is vital that the contractors and specialists who undertake any work are suitably qualified and experienced in dealing with historic buildings. In order to address this particular issue, under the aegis of

Historic Scotland and English Heritage, and in consultation with various professional institutions, a scheme for accreditation of conservation skills has been proposed and is currently being implemented. The aim is to provide a list of suitably qualified and recognised individuals who have been accredited by their professional bodies so that clients might be assured of the competence of their professionals.

The Architects Accredited in Building Conservation register lists those architects who have been independently accredited as suitably qualified to work in conservation. The Royal Institution of Chartered Surveyors also has a register of surveyors who are accredited in conservation. This scheme was set up in 1992 and aims to promote good practice within conservation. Other institutions are currently formulating an accreditation process for providing clients with a list of appropriately experienced practitioners. Amongst these are the recently chartered Institute of Architectural Technologists, whose scheme for accreditation is currently being implemented.

A similar evaluation of skill and competence must be adopted when choosing contractors. Only contractors with the necessary skills and experience of working on conservation projects should be appointed to deal with work on a historic building.

In addition to appointing suitable professionals and contractors, it is also vital that the choice of contractual agreement is appropriate in order to ensure control not only of quality but also safety and protection of what is a vulnerable asset. It needs to be remembered that standard forms of contract may not be appropriate or may require modification to reflect the special requirements when working on historic buildings.

Certain types of work or work methods which may be acceptable in conventional building work may be wholly inappropriate for historic buildings. The use of hot working methods is a prime example. Numerous fires have occurred in historic buildings as a result of this, including at Hampton Court in 1986, at Uppark House during roofing work in 1989 and at St George's Hall, Windsor during refurbishment in 1992. In these examples the use of hot methods of working resulted in major fires due to lack of control and insufficient monitoring of fire-risk construction methods.

In order to prevent such devastating fires a process of checks and daily inspections of vulnerable work areas must be implemented and maintained when hazardous operations are required. If at all possible, hazardous work methods must be avoided and suitable alternatives adopted. A risk assessment may be useful in assessing what may or may not be suitable. Where risky or hazardous work is inevitable, it may be necessary to undertake such work in an isolated area away from the building and then reintroduce the completed component once the hazardous operation is finished. Daily inspection of hazardous work areas must be undertaken scrupulously at the end of each working day.

Certain buildings may require specialist protection measures to prevent damage to fragile or vulnerable elements. This may include simple protection such as against dust, or may involve climate control or protection of vulnerable floor areas where tradesmen will be walking. Special precautions are inevitable and must be designed and maintained appropriately. Contractual arrangements must reflect these specialist needs and all personnel involved in the site must be aware of the particular risks and hazards for that location. Even if they have worked on other historic buildings, there may be new or special circumstances as every historic building is unique and requires appropriate consideration. Site staff may need to be specially trained and educated on the risks inherent – these must be defined in any contractual arrangements along with specification of methods of working.

Record works on completion

Once a comprehensive investigation has been undertaken into a site or building, it is important to archive any acquired information. The data may include details of the history of the building, information relating to previous projects and its patterns of deterioration and decay. The availability of such a record can help to inform future decision making.

It is also vital that any interventions undertaken are carefully recorded and stored with this information. Accurate recording must be an inherent part of the process as this will assist professionals when they are subsequently making decisions.

The depositing of acquired information must be carefully considered so that it is accessible in the future. Some buildings may have their own library which might prove suitable, but the majority of sites will not have such a facility and an alternative must be sought. Such facilities might be provided by the following:

- local reference libraries
- local records offices
- local authority records (SMRs)
- national archives and libraries
- English Heritage, Historic Scotland, CADW.

The act of recording, post-intervention, completes the cycle of conservation response but is as important a part of it as any other element within the cycle.

Chapter 49

Dealing with historic buildings in practice

In dealing with the historic environment, there may be buildings or sites that present particular issues or challenges. In these cases careful investigation and a conservation plan are appropriate.

Redundant buildings and change of use

Redundancy of historic buildings is a considerable concern to both conservation professionals and to those concerned with townscapes and urban prosperity. If a building becomes empty it can very quickly deteriorate. Lack of heat or maintenance and increasing risk of vandalism and arson are all threats. Even more minor problems such as pigeon infestation or growth of plants in gutters can eventually cause serious problems to a building. It is therefore vital that a building finds an economically viable re-use and in some cases this may prove to be a challenge.

Some owners may not have the desire for a building to be restored. Unless the building is listed or in a conservation area protective or enforcement action cannot be taken by a local authority. Even then enforcement can be a lengthy and expensive process that is not undertaken lightly by the relevant authorities unless absolutely necessary. There are also cases of building owners demolishing listed buildings without consent rather than repairing them, such as Greenside adjacent to the Wentworth golf course. Greenside was a 1937 Modernist building designed by Colin Lucas.

Historic structures provide a physical connection with the past. Some of these buildings will have a viable re-use but others will remain a link with the past, forming a physical record of former society and construction methods. For example, a doocot (dovecot) is not a

49.1   Doocot in Crail, Fife

building that can usefully be used as its original purpose has long since been superseded. Also, because of its construction style, small and with no windows, it really has no alternative use. The possibilities for such structures are really only as attractive buildings in the landscape, but that is not to denigrate that value. They make an important contribution in terms of style, aesthetic and historical record of former agricultural practices and rural living. However, they will still require maintenance and care if they are not to seriously deteriorate.

However, the majority of buildings need to find an appropriate and viable re-use. Such adaptation provides a sustainable resource in an increasingly ephemeral world where natural resources are now recognised as being finite. As historic buildings become redundant, it is necessary to find a suitable alternative use as without this and as empty buildings, they quickly deteriorate. The original use for which a building was designed is generally the best possible use. However, with shifts in economic opportunities, locations change in emphasis and it becomes unrealistic to maintain the original use. Suitable alternatives will vary depending on the nature and layout of the building. Some may lend themselves to different uses while others, by their location, layout or perhaps size, are inflexible and difficult to convert to an alternative viable use.

Buildings under former agricultural use that have become redundant through changes in farming practices have probably been subject to extended periods of decline of those uses. During such periods of reducing use wildlife such as bats, barn owls and other rare species may have taken the opportunity of residence as these buildings provide ideal habitats as human use declines. Once a decision is made to convert these buildings to alternative uses the wildlife inhabitants and their habitat come under severe threat.

The Wildlife and Countryside Act 1981 was implemented in order to provide protection and to constrain actions which might threaten wildlife or their habitats. It is complimented by the Conservation (natural habitat, etc.) Regulations 1994. Both of these pieces of regulatory control have a focus on ensuring that action such as redevelopment or conversion of buildings impacts as little as possible on the established habitats of wildlife or their patterns of use.

It is very much a part of the conversion and investigation process of redundant buildings, prior to implementation of intervention works, that a full and detailed study and impact analysis of wildlife usage is made. The conversion work programme and phasing may need to be adjusted to suit breeding patterns of rare species such as bats, for example, and alternative accommodation may need to be provided for incumbent owls. Much guidance is available from Natural England on reduction of impact on wildlife by conversion of redundant structure and DEFRA are empowered to issue licences so that only approved personnel may investigate or interfere with bat habitats.

Any investigation plan or intervention plan must take cognizance of the imposed legislation and incorporate its provisions in any investigation work.

Increasingly there is demand for conversion to domestic use. With steep rises in house prices and high demand for houses, such a development can be financially lucrative and therefore attractive. However, such conversions can be fraught with difficulties, especially if the building is listed and/or is inflexible. Barns are particularly popular conversions to residential in rural areas. However, it can be difficult to achieve this successfully and to marry the original style and functionality of the barn with the requirements of a domestic residence.

Cunnington (1988) states that although redundant farm buildings can offer considerable scope for conversion, this needs to be carried out very sympathetically if their simple functional character is not to be lost. This applies not only to the building but also to their setting. For example, it is important to retain elements such as a sack hoist. This appreciation of a building's setting is crucial to its viability and to its sensitive re-use. In order to achieve the best outcome for a building the most important aspect is to fully understand its setting and location. Armed with full knowledge about the history and development of a building it is much easier to fully consider suitable compatible uses. A full survey and historical investigation are therefore necessary as a

49.2   Gladstone Pottery museum, Stoke-on-Trent

starting point for conversion. This initial work can help to prevent problems arising during the development by anticipating potential issues and conflicts.

The end result of such thorough investigation can be a successful scheme. This was the case for Perth Waterworks in Perth, Scotland. The cast-iron waterworks was built in 1832 by Dr Adam Anderson, Rector of Perth Academy, in order to provide the town with a clean water supply. Its status as one of the earliest cast-iron buildings is reflected in its listing as a Category A building. It closed as a waterworks in 1965 and was subsequently used as a tourist office. However, it needed considerable repairs and this resulted in a major restoration project in 2003. The building has housed the Fergusson collection of paintings since 1992 and, with the new repairs, is a very successful art gallery displaying a collection of paintings by John Duncan Fergusson (1874–1961), one of Scotland's most important Colourist artists. This demonstrates how an unusual building can be successfully re-used.

49.3   Perth Waterworks, now used for the Fergusson gallery

It is therefore crucial to consider the long-term viability of a building when deciding on possible alternative uses. While the immediate reaction is to convert a historic building into a museum or heritage centre, there is a limit to how many of these an area can realistically sustain. Although such buildings may have a future, particularly in popular tourist locations, it is vital to remember that the building must be economically viable. Alternative use must be inventive and creative as well as sympathetic. The impact of visitor numbers can have a destructive effect on both the building and its setting. Concern is being expressed in respect of international sites such as the Parthenon in Greece and Matu Pitchu in Peru where visitor numbers are seriously affecting the longevity of the asset and its context. This both detracts from the building itself and from the visitor experience. With the availability of cheap flights across the globe, tourism is booming and historic centres like Florence, Edinburgh, Rome, Bath and Athens often struggle to cope. The balance between economics and conservation can be a difficult one that is not easily resolved.

Certain buildings struggle to find a suitable alternative use, especially those with awkward sites or challenging internal layouts. Churches can be particularly problematic, as are large industrial mills and also cinemas. Our towns are littered with the remains of many derelict 1930s cinemas which have been upstaged by new multiplexes located on the edge of towns with dedicated parking. Some had a new life as bingo halls, but even this use is in decline and increasingly permission

49.4   Tourists in Florence

49.5 Unsympathetic alterations to a church

49.6 The Hub, Edinburgh

for demolition and redevelopment of the site is being sought, particularly where they are not listed. The saving of such buildings is often initiated by a local community who are anxious to save part of their town's history. These may also form part of grant schemes such as Townscape Heritage Initiatives (THIs) and Conservation Area Regeneration Schemes (CARs) where grant-funded regeneration is used to improve economically deprived areas.

Churches have strong emotional connections with an area, even more so where there is a graveyard surrounding the church. In some cases there is no viable alternative for a church, especially where it sits remotely in a rural area. In this case there are societies such as the Scottish Redundant Churches Trust and the Friends of Friendless Churches who may acquire them and ensure that they remain wind- and water-tight. In urban areas there may be more alternatives for re-use. Some of these may be retail such as a furniture warehouse, but in other cases the use may be as a community centre or as leisure space. The Hub in Edinburgh's Royal Mile is an example of the use of a former assembly hall for the Church of Scotland for a new purpose. It was designed by architects James Gillespie Graham and Augustus Pugin in the 1840s but was acquired by the Edinburgh Festival in 1995 and was renovated to provide a very successful venue for the international festival held in Edinburgh every year, as well as other events during the year.

Façade retention

In some cases there may be a proposal to retain the original façade of a building but to build a completely new structure behind that. This gives a building a prestigious façade but behind is a modern building that can accommodate all the needs of its occupiers. The increased use of the site may be more attractive to developers and make it more economically viable. While outwardly the building may appear historic, this is effectively an architectural deceit. However, the economic pressures for sites to maximise their income can sometimes overtake the desire to retain historic fabric.

Highfield (1991) suggests that there is acceptance of some of these façade-retention schemes, because, it is argued, they still maintain the 'townscape' element of the building which is the part which forms the public function and is not necessarily either historically or architecturally related to its interior. However, although it may be considered as the only means of achieving an economically viable re-use for many historic buildings, faced-retention schemes should only be undertaken when all other options have been exhausted. As Richards (1994: 2) states: "There is also criticism that facadism prevents new architectural styles from evolving, and reduces the design of buildings to mere two-dimensional elevations 'creating townscapes which are little more than stage sets'."

Ruins and ancient monuments: their preservation and protection

Once a building falls out of use, over a considerable period of time both structure and fabric can deteriorate to a point where renovation or restoration becomes impossible. Nonetheless, these ruined structures are part of our cultural and historic heritage and still have a part to play in representing our history and development. Their preservation and repair require a different approach to conservation and use of buildings that still have a useful life. Likewise, ancient monuments, such as stone circles or standing stones, will also require particular care that may differ from that taken to an inhabited building.

While this part is concerned with conservation rather than preservation, the approach required for ruins differs. It is appropriate to consider the definitions of these terms as outlined in BS 7913:

> **Conservation**: Action to ensure the survival or preservation of buildings, cultural artefacts, natural resources, energy or any other thing of acknowledged value for the future.

Preservation: State of survival of a building or artefact, whether by historical accident or through a combination of protection and active conservation.

Perhaps the easiest way to define the difference between conservation and preservation is that the former may involve activities that ensure the survival of the artefact by a continuum of appropriate use or re-use and the latter by consolidation and protection of the artefact by minimal protective intervention against further deterioration.

John Ruskin addressed the subject in his writings for the SPAB and, particularly in *The Seven Lamps of Architecture*, made recommendations that are still relevant today. He states that we should

> call upon those who have to deal with them to put Protection in the place of Restoration, to stave off decay by daily care, to prop a perilous wall or mend a leaky roof by such means as are obviously meant for support or covering and show no pretence of other art.

The important definition by Ruskin being in the words "showing no pretence of other art", in other words do no more than is necessary to keep the wind and weather out in order to preserve the ruined heritage as it stands.

A medieval ruined abbey is just as important a reminder of heritage as a restored and working watermill. The protection and perpetuation of the ruin is equally valid as heritage as a cotton mill conserved as a working museum. However, they present some particular problems in trying to maintain that ruined state. A building which is ruinous will naturally continue to degenerate into a pile of stones which are eventually unrecognisable as a building. Ruins therefore need to be consolidated in order that they are maintained, but this effectively interferes with that natural process making this a difficult balancing act. The fact that such buildings generally do not have a roof means that wallheads need to be consolidated with suitable material, sometimes turf or mortar. Stone or brickwork needs pointing to prevent it from collapsing and vegetation must be controlled. However, to some extent the presence of vegetation adds to the romantic presence of such buildings and may also be home for particular flora and fauna such as lichens, insects and rare plants. Again, the control of vegetation is a delicate balance between protecting the structure, protecting wildlife and ensuring the aesthetic value of the building. This can be seen at Fountains Abbey in Yorkshire where the ruins of the abbey are carefully maintained so they retain their picturesque landscape and do not deteriorate.

There is now a greater understanding of ruins than in the past. Works of major alteration were carried out that perhaps would not be done today. For example, Eilean Donan Castle on the west coast of Scotland

49.7   Ruined wall of a church capped with turf

is an iconic image of a Scottish castle. However, it was destroyed during the Jacobite uprisings in 1719 and was then ruinous until rebuilt in the early twentieth century. Whether such a major rebuilding would be done today would take considerable debate as to the most appropriate course of action for the building given current conservation philosophy.

In recent years there has been increasing interest in ancient monument sites such as standing-stone circles. While major sites like Stonehenge have always created interest, some of the more remote stones are now increasing in popularity. The Ring of Brodgar in Orkney and Callanish Stones in Lewis do not have the same pressures as Stonehenge, but all these sites need careful management. Part of the experience for visitors is the ancient feel of such a site and the connection with prehistoric civilisations. That can be hard to achieve where visitor numbers are high. At Stonehenge, to protect the stones, visitors have been prevented from getting near to the stones themselves and instead they have to walk around a remote walkway, making for a disappointing visit for some. However, English Heritage who manage this World Heritage site have reappraised the site through a management plan which has examined access and other impacts on the site, such as the close proximity of a main road which also detracts from the visitor experience. As part of this, improvements to the site will include the removal or tunnelling of roads so they do not impact on the site. More remote sites like a broch in

49.8   Ruins of Belchite in Northern Spain, blitz-bombed by Franco forces during the Spanish Civil War of the 1930s and left as a reminder of that event

49.9   Eilean Donan Castle, Scotland

highland Scotland with low visitor numbers will require less management and potentially intervention in the site.

More remote sites also need to be carefully managed so that visitors do not damage the site through excessive footfall. Signage and visitor interpretation needs to be carefully thought out so it is informative yet not intrusive into the site itself. There may also need to be some more major intervention in order to protect an ancient monument. The Pictish Sueno Stone in Forres, Moray, Scotland is over six metres high and dates to the first millennium AD. It is considered so vulnerable that a protective glass box has been put around the stone. Such a method of protection is never taken lightly and all suitable options will be considered. Inevitably this type of protection will interfere with the aesthetic appearance of a monument, but this has to be balanced with the long-term future of such an important structure. In other cases, a stone of importance may be moved to a suitable place indoors where the weathering effects will be minimal. The ninth-century Pictish Dupplin Cross was originally sited in a field near Dunning in Perthshire. It was moved to the National Museum of Scotland in Edinburgh but then returned to the village and housed in St Serf's Church, a property in the care of Historic Scotland. While not in its original position outside the village, the location within the church is a compromise that both protects the stone, satisfies the desire of the community to have it at its original location and allows visitors to

49.10   A broch, Scotland

49.11   Sueno Stone, Forres, Scotland

view it. Inevitably discussions and compromises need to be made for satisfactory resolutions to be arrived at.

In dealing with ruins and ancient monuments, conservation, preservation and archaeology overlap and we need to remind ourselves that these three elements are interdependent in understanding heritage and the conservation and preservation of it.

Chapter **50**

Sustainability and climate change

Conservation practitioners have a duty to consider the environment and the sustainability of projects. With increasing concerns about climate change and the depletion of valuable natural resources, conservation can play a key role in offering an eminently sustainable alternative to modern development.

Historic buildings contain considerable embodied energy through the materials which they contain and the human effort to build them. It took considerable energy to quarry stone or make bricks and to produce the lime that bonds them. To demolish such buildings and use their remains as landfill or hardcore is not a sustainable alternative. Instead, historic buildings should be regarded as important because of what they represent in terms of sustainable alternatives to new development.

However, climate change has direct implications for historic buildings in that increased rainfall may cause considerable problems. Drainage systems may be unable to cope with unusual and more frequent events such as flooding. The increasing frequency and magnitude of gale-force winds are potentially damaging to historic fabric. The conservation practitioner needs to be aware of these changes and to keep up to date with current thinking in this field.

Coastal erosion, by way of example and, potentially, a product of global warming, might force on conservators a decision structure that might, otherwise, not be a usual consideration. Coastal heritage sites where land is subject to loss through erosion and rising sea levels is a contextual condition that inland, high-ground sites are not subject to. Even then, non-coastal sites will be subject to an increasing rapidity of change induced by climate change. Such effects might be:

- increased rainfall and water penetration; resulting in
- increases in ambient moisture and temperature-creating situations

that buildings were not originally subject to – thermal shock, increases in moisture, etc., all of which will materially alter the normal equilibrium of a building;

● increases in drying and wetting rates due to increase in temperature and moisture to which an asset may not have previously been exposed.

Previously stable materials in a formerly stable environmental state may become subject to climatic changes to which they will not previously have had to adapt. The same analogy might be drawn with buildings that are subject to a new use in order to maintain an economic future. It is the impact of such changes that the conservators must evaluate in order to make a best possible scenario approach to change.

Chapter **51**

Conclusion

This part has considered the application of philosophy to historic buildings and the central role that conservation plans play in protecting the historic environment. It is essential to fully investigate a building before carrying out any work and to then apply the appropriate philosophical conservation approach of minimal intervention and respect for the cultural significance of a place.

This is particularly important with buildings which may present particular challenges. Certain buildings may become redundant and difficult to re-use. Other sites may have large visitor numbers and the sheer volume of visitors may be putting the quality and protection of these buildings under threat.

Careful investigation will inform of the vulnerability of a site so that a suitable management plan can be put in place to ensure the continued survival of these buildings.

Part 8
Overview

If history records good things of good men, the thoughtful
hearer is encouraged to imitate what is good.

The Venerable Bede, AD 673–735

In Part 1, we referred to the quote by Fielden, "conservation is about
controlling change". This book has examined three main inter-related
topics, architecture, materials and conservation. These all reflect a
continuum of social change and evolution. Central to this is an understanding that change naturally occurs as building owners strive to
make their properties perform the tasks for which they bought or
erected them. Sometimes buildings can be difficult to adapt and problems in their use or viability arise. Society revises its focus in terms of
fashion, needs and even location. Buildings may therefore become
redundant or simply go out of fashion. This has, in the past, resulted
in the destruction of many valuable buildings which would otherwise
have had a useful life.

However, with hindsight and a sharper focus on conservation, it can
be seen that buildings have embodied energy and the perpetuation of
a building is therefore a much more sustainable approach than demolishing it and rebuilding on the site. While economics may dictate this
provides the best option, gradually it is being realised that such an
approach cannot continue. Change can be difficult to accommodate in
a historic building. That is why it is essential to have an appropriate
philosophy to control or manage *appropriate* change.

Understanding architectural development is the basis for designing,
adapting and conserving buildings. The buildings erected by the
Ancient Egyptians, Greeks and Romans are the foundation of later
styles in architecture, including the Classical buildings of the eighteenth
and nineteenth centuries, as well as later revival styles found in the
Victorian and Edwardian periods. As society has adapted and evolved,

the needs of their buildings have changed. Economic circumstances have played a key part in architecture, but other influences include the political stability of a region, the influence of trade routes and the particular geography and geology of a region. Architecture also developed along plagiaristic lines, taking influence from what had gone before while, at the same time, adapting previous influence to social and fashionable changes in styles. These inter-related factors form a complex mixture of elements which alter their emphasis over time. However, studying buildings of all types and eras helps to provide an understanding of these elements. There is no substitute for looking at standing buildings to see how they are constructed and being able to study the materials and construction methods that formed them.

The understanding of building materials is of great importance. The choice of materials used to construct a building fundamentally affect its aesthetics, its design and its longevity. It will also affect the maintenance regime required and the potential for alteration or change. Some buildings may be quite robust in design but others, such as earth structures, may be much more vulnerable if not cared for properly. That said, all buildings require maintenance and repair and the use of appropriate materials is a crucial element of such work. The extensive use of cementatious mortars and renders in historic buildings during the twentieth century has had a serious and detrimental effect, causing erosion of bricks, stone and earth buildings. Considerable knowledge on the use of lime has been lost and the implications of this for the historic environment have only recently been recognised. Action is now being taken to improve training and to develop the appropriate knowledge and skills base necessary, but this remains an on-going area of research.

While considerable research has been undertaken into architectural history, materials and design, there are new challenges which threaten the future of historic buildings. The loss of skills in the use of lime is just one aspect of change in the building industry which has occurred during the twentieth century. The move to modern, rapid-construction techniques meant that traditional building skills in stone-masonry, iron-founding, brick-making and joinery have all been depleted. The availability of apprenticeship schemes has been limited and the result is that there are many buildings that are in serious disrepair. Without the appropriate skills and often the materials to repair them, they remain at risk. While this issue is being addressed by heritage organisations like Historic Scotland and English Heritage together with organisations like CITB and COTAC and building craft colleges, it will take time and money to train a critical mass of appropriately trained and qualified personnel.

The on-going threat of climate change is another area which requires further research in order that the threat to historic buildings is mitigated. Increased rainfall and storms mean that buildings may be vulnerable. This is an area where building owners will need to become increasingly aware. Coastal erosion, also linked to climate change, is an additional area where heritage structures are placed at risk and new methods will need to be developed to deal with such threats. This might include relocation of buildings away from threats posed, which has an important impact on context and setting which is as important a facet of understanding as the structure itself.

If we are to value our built environment we need to understand it, how and why it was formed and how it needs to be conserved, not only in respect for the past, but also for the benefit of future generations. We are mere custodians or stewards of it and we have a duty to use it appropriately and to pass it on to future generations to appreciate, but also to allow them to form their own values of it. Being able to pass this on requires adopting a suitable approach to its conservation. These approaches are set out clearly in international charters which outline the correct philosophical approaches to be adopted. These international charters have become a respected source of guidance over the years of their development and promulgation and, while it is recognised that they are an important source of guidance, they cannot provide a panacea or lexicon of acceptable response. Each project is individual and requires a bespoke response based on clarity of understanding of its peculiar and unique circumstances and use of recognised principles of conservation philosophy and practice developed through discussion and consensus.

Although some buildings may be so precious to us that they are conserved in a museum-like state, the vast majority will need to adapt and to be capable of viable economic use. Otherwise they become redundant and will deteriorate. The built environment therefore needs to be adaptable to our current needs if it is to survive at all. What we must not do is expect it to be able to conform to our current requirements without a detailed study of it, what it represents and how it may accept change. This is not to imply that the built environment must never change; for that would be preservation and that is not what conservation advocates. This is where conservation and preservation differ in their particular approach. Preservation is usually translated as the perpetuation of state at a point in time in order to preserve that state for future reference.

The built environment is subject to a continuum of change and adaptation throughout its existence. This demonstrates the value of the built environment as a palimpsest of history – a true physical record. Conservation, by comparison with preservation, is valuing the built environment for what it offers both as a resource for understanding history as well as a resource for use. However, the change to facilitate its re-use must be appropriate and made within the ability of an asset to be adaptive of change without loss of significance and record. Conservation is about managing appropriate change in a way that does not threaten significance and record. We must value our built heritage for what it is, what it displays and what it is capable of becoming. Recognition of the limitations of a building is derived from a total knowledge of what it is, what its capacity for change is and what it is not able to accomplish without loss of its intrinsic aesthetic value or cultural significance.

Buildings, both historic and modern, are central to people's lives. Their design and aesthetics are influenced by the way people live and work and vice versa. Both historic townscapes and rural landscapes are part of history and their continued use is the future. Buildings must therefore be respected, maintained and used under the stewardship of current generations so that future generations can appreciate the importance of the past.

> We shape our buildings and afterwards our buildings shape us.
>
> (Winston S. Churchill)

Bibliography

Adams, I. (1978). *The Making of Urban Scotland*. London, Croom Helm Ltd.

Adams, S. (1987). *The Arts & Crafts Movement*. Herts, The Apple Press Ltd.

Alcock, N. W. with M. W. Barley, P. W. Dixon and R. A. Meeson (1996). *Recording Timber Framed Buildings: An Illustrated Glossary*. London, Council for British Archaeology.

Arnold, D. (2002). *Reading Architectural History*. London, Routledge.

Ashurst, J. (1998). *Practical Building Conservation Series: Volume 4 Metals*. Aldershot, Gower.

Ashurst, J. and J. Ashurst (1988). *Practical Building Conservation: Volume 2 Brick Terracotta and Earth*. Aldershot, Gower.

Ashurst, J. and N. Ashurst (1996). *Practical Building Conservation: Volume 3 Plasters, Mortars and Renders*. Aldershot, Gower.

Aston, M. and J. Bond (1976). *The Landscape of Towns*. London, JM Dent.

Bailey, H. and D. Hancock (1990). *Brickwork and Associated Studies Vol 2*. London, Macmillan.

Ballantine, I. (1997). *Cast Iron. Traditional Building Materials*. Edinburgh, Historic Scotland.

Bannister, T. (1950). "The first iron framed buildings", *Architectural Review,* April: 231–246.

Beaton, E. (1997). *Scotland's Traditional Houses: From Cottage to Tower House*. Edinburgh, The Stationary Office.

Bell, D. (1997). *Technical Advice Note 8: The Historic Scotland*

Guide to International Conservation Charters. Edinburgh, Historic Scotland.

Bell, D. (2001). *Research into Structured Support for CPD Development*. Draft report commissioned by the School of Architecture, Edinburgh College of Art, in co-operation with Historic Scotland and Heriot-Watt University's Department of Building Engineering and Surveying (now School of the Built Environment).

Bennett, B. (2001). "Awash with colour: the use of limewash as a decorative and protective coating", *The Building Conservation Directory*. Online, available at: www.buildingconservation.com/articles/awash/awash.html.

Boulting, N. (1976). "The law's delays: conservationist legislation in the British Isles", in Fawcett, J. (ed). *The Future of the Past: Attitudes to Conservation 1174–1974*. London, Thames & Hudson.

Brereton, C. (1995). *The Repair of Historic Buildings: Advice on Principles and Methods*. London, English Heritage.

Brown, R. J. (1982a). *The English Country Cottage*. London, Robert Hale.

Brown, R. J. (1982b). *English Farmhouses*. London, Robert Hale.

Brunskill, R. W. (1981). *Traditional Buildings of Britain*. London, Victor Gollancz.

Brunskill, R. W. (1982). *Illustrated Handbook of Vernacular Architecture*. London, Faber & Faber.

Brunskill, R. W. (1999). *Traditional Farm Buildings of Britain and Their Conservation*. London, Victor Gollancz.

Centre for Conservation and Urban Studies (CCUS) (2000). *Scottish Slate: The Potential for Use in Building Repair and Conservation Area Enhancement*. Edinburgh, Historic Scotland.

Charles, F. W. B. with M. Charles (1986). *Conservation of Timber Buildings*. Shaftesbury, Donhead Publishing.

Clark, K. (1998). *Conservation Plans in Action. Conservation Plans for Historic Places*. Oxford, English Heritage.

Clark, K. (2001). *Informed Conservation: Understanding Buildings and Their Landscapes for Conservation*. London, English Heritage.

Clifton-Taylor, A. (1987). *The Pattern of English Building*. London, Faber & Faber.

Collings, J. (2002). *Old House Care and Repair*. Shaftesbury, Donhead Publishing.

Constantinides, I. and L. Humphries (2003). "Exterior stucco", *The*

Building Conservation Directory. Online, available at: www.building conservation.com/articles/stucco/stucco.htm.

Conway, H. and R. Roenisch (2005). *Understanding Architecture: An Introduction to Architecture and Architectural History*. London, Routledge.

Cooper, J. (ed.) (1989). *Mackintosh Architecture: The Complete Buildings and Selected Projects*. New York, Academy Editions.

Cox, A. and I. Thomson (1998). *Contracting for Business Success*. London, Thomas Telford.

Cox, A. and M. Townsend (1998). *Strategic Procurement in Construction*. London, Thomas Telford.

Cranfield, I. (2002). *Georgian House Style*. Newton Abbot, David & Charles.

Cruickshank, D. and P. Wyld (1977). *London: The Art of Georgian Building*. London, Architectural Press.

Cullingworth, J. B. and V. Nadin (1997). *Town and Country Planning in the UK*. London, Routledge.

Cunnington, P. (1988). *Change of Use*. London, A & C Black.

Curl, J. S. (1990). *Victorian Architecture*. London, David & Charles.

Curl, J. S. (1993). *Georgian Architecture*. London, David & Charles.

Curl, J. S. (1999). *Oxford Dictionary of Architecture*. Oxford, Oxford University Press.

Dallas, R. (ed.) (2003). *Guide for Practitioners 4: Measured Surveys and Building Recording*. Edinburgh, Historic Scotland.

Davey, A., B. Heath, D. Hodges, R. Milne and M. Palmer (1995). *The Care & Conservation of Georgian Houses: A Maintenance Manual*. Oxford, Butterworth Architecture.

Department of Culture, Media and Sport (DCOMS) (2007). *Heritage Protection for the 21st Century*. London, HMSO.

Dunbar, J. G. (1966). *The Historic Architecture of Scotland*. London, BT Batsford.

Dunbar, J. G. (1978). *The Architecture of Scotland*. London, BT Batsford.

Duncan, A. (1988). *Art Deco*. London, Thames & Hudson.

Earl, J. (1997). *Building Conservation Philosophy*. Reading, College of Estate Management.

Eldridge, M. (1958). "The plate glass shop front", *Architectural Review*, 123: 192–195.

English Heritage (2000). *Power of Place: The Future of the Historic Environment*. London, English Heritage.

English Heritage (2006). *History Matters: Pass it On*. London, English Heritage.

Fawcett, R. (1994). *Scottish Architecture from the Accession of the Stewarts to the Reformation*. Edinburgh, Edinburgh University Press.

Fellows, R. (1995). *Edwardian Architecture: Style and Technology*. London, Lund Humphries.

Fenton, A. and B. Walker (1981). *The Rural Architecture of Scotland*. Edinburgh, John Donald Publishers Ltd.

Fenwick, H. (1974). *Scotland's Historic Buildings*. London, Robert Hale.

Fletcher, Banister (1928). *A History of Architecture on the Comparative Method*. London, Batsford.

Furneaux Jordan, R. (1997). *Western Architecture*. London, Thames & Hudson.

Gloag, J. and D. Bridgewater (1948). *A History of Cast Iron in Architecture*. London, George Allen and Unwin Ltd.

Goode, W. J. (1994). *Round Tower Churches of South East England*. Burnham Market, Round Tower Church Society.

Gray, A. (1985). *Edwardian Architecture: A Biographical Dictionary*, London, Duckworth.

Grigg, J. (1987). *Charles Rennie Mackintosh*. Glasgow, Richard Drew Publishing.

Haynes, N. (2000). *Perth and Kinross: An Illustrated Architectural Guide*. Edinburgh, Rutland Press.

Hayward Gallery (1988). *Lutyens: The work of the English Architect Sir Edwin Lutyens (1869–1944)*. London, Arts Council of Great Britain.

Heritage Lottery Fund (1998). *Conservation Plans for Historic Places*. London, Heritage Lottery Fund.

Herman, A. (2003). *The Scottish Enlightenment: The Scots' Invention of the Modern World*. London, Harper Perennial.

Hibbert, C. (1997). *Florence: The Biography of a City*. London, The Folio Society.

Highfield, D. (1991). *The Construction of New Buildings Behind Historic Facades*. London, E & F Spon.

Historic Scotland (1998a). *A Guide to Conservation Areas in Scotland*. Edinburgh, Historic Scotland.

Historic Scotland (1998b). *Memorandum of Guidance on Listed Buildings and Conservation Areas*. Edinburgh, Historic Scotland.

Historic Scotland (1998c). *Scotland's Listed Buildings: A Guide for Owners and Occupiers*. Edinburgh, Historic Scotland.

Historic Scotland (2000). *Conservation Plans: A Guide to the Preparation of Conservation Plans*. Edinburgh, Historic Scotland.

Historic Scotland (2001). *The Historical and Technical Development of the Sash and Case Window in Scotland*. Edinburgh, Historic Scotland.

HMSO (1989). *Defects in Buildings*. London, HMSO.

Hockman, H. (2002). *Edwardian House Style*. Newton Abbot, David & Charles.

Holder, J. (2002). "Invisible menders", *The Architects Journal*. Online, available at: www.architectsjournal.co.uk/archive/invisible_menders.html

Howard, D. (1995). *The Architectural History of Scotland: Scottish Architecture from the Reformation to the Restoration*. Edinburgh, Edinburgh University Press.

Howarth, T. (1977). *Charles Rennie Mackintosh and the Modern Movement*. Henley on Thames, Routledge.

Humberstone, J. (1997). "Taking the philosophical approach", *The Building Conservation Directory*. N.p, Cathedral Communications Limited.

Hume, J. R. (1992). "Iron in buildings in Scotland", in A. Riches and G. Stell (eds) *Materials and Traditions in Scottish Building*. Edinburgh, Scottish Vernacular Buildings Working Group.

ICOMOS (1990). *Guide to Recording Historic Buildings*. London, Butterworth Architecture.

Innocent, C. F. (1999). *The Development of English Building Construction*. Shaftesbury, Donhead Publishing.

Jamieson, F. (1993). *Drummond Castle Gardens*. The Grimsthorpe and Drummond Castle Trust.

Keefe, L. (2005). *Earth Buildings: Methods, Materials, Repair and Conservation*. Abingdon, Taylor and Francis.

Kent, R. (1999). "Ecclesiastical exemption: an update", *Building Conservation Directory*. Online, available at: www.buildingconservation.com/articles/exempt/exempt.htm.

Kerr, J. S. (1998). *The Conservation Plan. Conservation Plans in Action*. Oxford, English Heritage.

Knight, J. (1995). *The Repair of Historic Buildings in Scotland*. Edinburgh, Historic Scotland.

Koppelkamm, S. (1981). *Glasshouses & Wintergardens of the Nineteenth Century*. New York, Granada.

Lancaster, O. (1976). "What Should We Preserve?", in J. Fawcett (ed.) *The Future of the Past: Attitudes to Conservation, 1174–1974*. London, Thames & Hudson.

Latham, D. (2000). *Creative Re-use of Buildings*. Volumes 1 and 2. Shaftesbury, Donhead Publishing.

Lead Sheet Association (2008). *Properties of Lead Sheet*. Lead Sheet Association. Online, available at: www.leadsheetassociation.org.uk/html/1200.html.

Lynch, G. (1994). *Brickwork: History, Technology and Practice*. Volumes 1 and 2. Shaftesbury, Donhead Publishing.

Macdonald, S. (ed.) (1996). *Modern Matters: Principles and Practice in Conserving Recent Architecture*. Shaftesbury, Donhead Publishing.

Macdonald, S. (ed.) (2001). *Preserving Post-War Heritage: The Care and Conservation of Mid-twentieth Century Architecture*. Shaftesbury, Donhead Publishing.

Macdonald, S., K. Normandin and B. Kindred (2007). *Conservation of Modern Architecture*. Shaftesbury, Donhead Publishing.

McFadzean, R. (1979). *The Life and Work of Alexander Thomson*. London, Routledge and Kegan Paul.

McGrath, R. and A. Frost (1961). *Glass in Architecture and Decoration*. London, Architectural Press.

McKean, C. (1987). *The Scottish Thirties*. Edinburgh, Scottish Academic Press.

McWilliam, C. (1975). *Scottish Townscape*. London, Collins.

Marquis-Kyle, P. and M. Walker (1996). *The Illustrated Burra Charter*. Australia, ICOMOS/Australian Heritage Commission.

Martin, G. H. (1968). "The town as a palimpsest", in H. J. Dyos (ed.) *The Study of Urban History*. London, Edward Arnold.

Mercer, E. (1979). *English Vernacular Houses*. London, RCHM.

Montgomery-Massingberd, H. and C. S. Sykes (2000). *Great Houses of England & Wales*. London, Laurence King.

Morley, C. (n.d.). *Monograph on Round Towers*.

Morriss, R. K. (2001). *The Archaeology of Buildings*. Stroud, Tempus Publishing Ltd.

Naismith, R. J. (1985). *Buildings of the Scottish Countryside*. London, Victor Gollancz Ltd.

Naismith, R. J. (1989). *The Story of Scotland's Towns*. Edinburgh, n.p.

National Trust for Scotland (2004). *Newhailes*. Edinburgh, National Trust for Scotland.

Neumann, D. (1995). "Prismatic glass", in T. C. Jester (ed.) *Twentieth Century Building Materials: History and Conservation*. New York, McGraw Hill.

Nicholson, A. (1997). *Restoration: The Rebuilding of Windsor Castle*. London, Penguin Books.

Nicolson, N. (1978). *The National Trust Book of Great Houses of Britain*. London, Granada.

Osborne, R. (2006). *Civilisation: A New History of the Western World*. London, Pimlico.

Pearson, L. (2005). *Tile Gazetteer*. London, Richard Dennis.

Penoyre, J. and J. Penoyre (1978). *Houses in the Landscape: A Regional Study of Vernacular Building Styles in England and Wales*. London, Faber & Faber.

Pevsner, N. (1966). *An Outline of European Architecture*. Harmondsworth, Penguin.

Pevsner, N. (1976). "Scrape and anti-scrape", in J. Fawcett (ed.) *The Future of the Past: Attitudes to Conservation 1174–1974*. London, Thames & Hudson.

Pickard, R. D. (2000) cited by Bell, D. and thought to originate in *Management of Historic Centres*. London, Spon.

Pilcher, D. (1947). *The Regency Style 1800 to 1830*. London, BT Batsford Ltd.

Powys, A. R. (1929). *Repair of Ancient Buildings*. London, SPAB.

Ramsey, C. R. and J. D. M. Harvey (1977). *Small Georgian Houses and their Details 1750–1820*. London, Architectural Press.

Reen, K. (1999). "Case study: the Old Schoolhouse, Cottown, Perthshire", *Context*, 63. Online, available at www.ihbc.org.uk/context_archive/63/schoolhouse/cottown.html.

Reid, R. (1980). *The Book of Buildings: A Traveller's Guide*. London, Michael Joseph.

Rice, M. (2006). *Village Buildings of Britain*. London, Time Warner.

Richards, J. (1994). *Facadism*. Abingdon, Taylor & Francis.

Richardson, B. A. (2001). *Defects and Deterioration in Buildings*. London, E & F Spon.

Ridout, B. (2001). *Timber Decay in Buildings: The Conservation Approach to Treatment*. London, E & F Spon.

Robson, P. (1999). *Structural Repair of Traditional Buildings*. Shaftesbury, Donhead Publishing.

Rodgers, N. (2006). *Ancient Rome*. London, Hermes House.

Roth. L. (1993). *Understanding Architecture*. London, Herbert Press.

Ruskin, J. (1849). *The Seven Lamps of Architecture*. N.p., J. Wiley.

Ruskin, J. (1851). *The Stones of Venice*. N.p., Elder.

Scottish Lime Centre Trust (1995). *Technical Advice Note 1: Preparation and Use of Lime Mortars*. Edinburgh, Historic Scotland.

Service, A. (ed.) (1975). *Edwardian Architecture and its Origins*. London, Architectural Press.

Service, A. (1977). *Edwardian Architecture: A Handbook to Building Design in Britain 1890–1914*. London, Thames & Hudson.

Silverman, D. P. (ed.) (1997). *Ancient Egypt*. London, Duncan Baird.

Sinclair, S. J. (ed.) (1977). *The Statistical Accounts for Scotland Vol XI North & North-West Perthshire 1791–1799*. Wakefield, EP Publishing Ltd.

Smith, Annette (1982). *Jacobite Estates and the '45*. Edinburgh, John Donald Publishers Ltd

Smith, L. (1985). *Investigating Old Buildings*. London, BT Batsford Ltd.

Stenning, D. F and D. D. Andrews (eds) (2002). *Regional Variations in Timber Framed Buildings in England and Wales down to 1550*. Chelmsford, Essex County Council.

Summerson, J. (1986). *The Architecture of the Eighteenth Century*. London, Thames & Hudson.

Summerson, J. (1996). *The Classical Language of Architecture*. London, Thames & Hudson.

Swailes, T. (2006). *Scottish Ironwork Structures*. Edinburgh, Historic Scotland.

Swallow, P., R. Dallas, S. Jackson and D. Watt (2004). *Measurement and Recording of Historic Buildings*. Shaftesbury, Donhead Publishing.

Tiedsell, S., T. Oc and T. Heath (1996). *Revitalising Historic Urban Quarters*. Oxford, Architectural Press.

Thomson, N. and P. Banfill (2005). "Corrugated-iron buildings: an endangered resource within the built heritage", *Journal of Architectural Conservation*, 1: 67–83.

Vicat, L. J. (1997). *A Practice and Scientific Treatise on Calcareous Mortars and Cements, Artificial and Natural*. Shaftesbury, Donhead Publishing.

Walker, B. and C. McGregor (1996a). *Earth Structures and Construction in Scotland*. Edinburgh, Historic Scotland.

Walker, B. and C. McGregor (1996b). *Technical Advice Note 5: The Hebridean Blackhouse*. Edinburgh: Historic Scotland.

Walker, B., C. McGregor and S. Gregor (1996). *Technical Advice Note 4: Thatch and Thatching Techniques*. Edinburgh, Historic Scotland.

Walsh, J. (2000). *Technical Advice Note 21: Scottish Slate Quarries*. Edinburgh, Historic Scotland.

Watkins, D. (1979). *English Architecture: A Concise History*. London, Thames & Hudson Ltd.

Watkins, D. (1982). *The Buildings of Britain: Regency*. London, Barrie and Jenkins.

Watts, M. (2000). *Wind and Water Power*. Princes Risborough, Shire Publications Ltd.

Weaver, L. (1987/1913). *Houses and Gardens by E.L. Lutyens*. London, Country Life.

Weaver, M. E. (1993). *Conserving Buildings: Guide to Techniques and Materials*. New York, John Wiley and Sons.

West, T. W. (1985). *Discovering Scottish Architecture*. Princes Risborough, Shire Publications Ltd.

Woodforde, J. (1976). *Bricks to Build a House*. London, Routledge.

Woodforde, J. (1985). *Georgian Houses for All*. London, Routledge and Kegan Paul.

Wright, A. (1991). *Techniques for Traditional Buildings*. London, Batsford.

Youngson, A. (1988). *The Making of Classical Edinburgh 1750–1840*. Edinburgh, Edinburgh University Press.

Index